THE GREENING OF CANADA
Federal Institutions and Decisions

Environmental matters have become increasingly important in Canadian and world policy agendas. In this study, G. Bruce Doern and Thomas Conway trace the development of Canadian environment policy, giving an in-depth account of twenty years of environmental politics, politicians, institutions, and decisions as seen through the evolution of Ottawa's lead policy agency, Environment Canada.

The Greening of Canada is an extensively researched look at the entire period from the early 1970s to the present and is the most complete and integrated analysis yet of federal environmental institutions and key decisions. From Great Lakes pollution to the Green Plan, from the Stockholm Conference to the post–Rio Earth Summit era, the authors deal with both domestic and international events and influences on Ottawa's often abortive efforts to entrench a green agenda into national politics.

The book explores the crucial relationships of institutional and political power, directing attention at the DOE and its parade of ministers, intra-cabinet battles, federal-provincial relations, business relations and public opinion, and international and Canada–U.S. relations. It also examines important topics from acid-rain policy to the politics of establishing national parks, and from the Green Plan to the realities of environmental enforcement. Employing a framework cast as the 'double dynamic' of environmental policy making, the authors show the growing struggle between the management of power among key institutions and the need to accommodate a biophysical realm characterized by increased uncertainty as well as scientific and technological controversy.

G. BRUCE DOERN is a professor in the School of Public Administration, Carleton University. He is the author and co-author of numerous books on Canadian politics and policy, including *Faith and Fear: The Free Trade Story*, with Brian Tomlin, and *Canadian Public Policy: Ideas, Structure, Process*, with Richard Phidd.

THOMAS CONWAY is a policy analyst and consultant in Ottawa. He has written extensively on environmental politics and regulation.

The Greening of Canada

Federal Institutions and Decisions

G. Bruce Doern and
Thomas Conway

UNIVERSITY OF TORONTO PRESS
Toronto Buffalo London

© University of Toronto Press Incorporated 1994
Toronto Buffalo London
Printed in Canada

ISBN 0-8020-0645-0 (cloth)
ISBN 0-8020-7599-1 (paper)

Printed on acid-free paper

Canadian Cataloguing in Publication Data

Doern, G. Bruce, 1942–
    The greening of Canada : federal institutions and decisions

    Includes bibliographical references and index.
    ISBN 0-8020-0645-0 (bound) ISBN 0-8020-7599-1 (pbk.)

    1. Environmental policy – Canada.   2. Canada.
    Environment Canada.   I. Conway, Thomas, 1957–
    II. Title.

    HC120.E5D6 1995   363.7'0560971   C94-931926-0

University of Toronto Press acknowledges the financial assistance to its
publishing program of the Canada Council and the Ontario Arts Council.

Contents

TABLES   viii

PREFACE   ix

ABBREVIATIONS   xi

Introduction   3

Key Questions and Issues   4
The Double Dynamic   6
The Political-Institutional Analysis of Environmental Policy   9
The Rise and Fall and Rise of the DOE   12

1 The Department of the Environment: A 'House Divided'   16

Trudeau and the 'Mandate Debates'   17
Statutory Capacity   21
Campaigning to Get Out: Fisheries, Forestry, and the Parks Swap   22
The Environmental Conserver and Protector Roles:
    Resource-Use Planning versus Regulation?   30
The Atmospheric Environment Service: From Service Provider
    to Global Researcher   33
Conclusions   36

2 Ministers, Mandarins, and the Green Agenda   38

Jack Davis and the Engineer   39
Blair Seaborn and the Liberal Group of Six   41
Early Conservative Ministers: Nowhere to Go But Up   46
Lucien Bouchard and the Green Plan Ministers   49

Departmental Planning and Learning   52
The Department and the Green Plan   56
Conclusions   58

3 **The DOE and the Ottawa System**   60

Searching for Central Support   61
Budgetary Battles   64
Other Departments   66
Scientific and Policy Knowledge as a Base of Influence   74
The Green Plan and the Ottawa System   78
Conclusions   81

4 **Environmental Federalism and Spatial Realities**   83

Water Pollution and Management: Conflict and Cooperation   85
Air Pollution: Institutionalized Cooperation   89
Managing General Environmental Federalism   91
Environmental Accords: The Need for Practical Bilateralism   95
The Provinces and the Green Plan   97
Conclusions   99

5 **The Elusive Constituency: ENGOs, Business, and the
   Public**   100

The ENGOs and the DOE: Reluctant Allies   101
Business and the DOE: From Inner Table to Round Table   107
Business–ENGO Convergence: New Maturity or New Battles?   111
Public Opinion: The Villain Is Us?   116
The Green Plan, the ENGOs, and Business   118
Conclusions   122

6 **International Environmental Relations**   124

Bilateral Relations and the Great Lakes Agreement   126
The United States as Environmental Hero and Villain   129
Multilateral Relations: Towards the '30 Per Cent Club'   132
The New Political Economy of Protocol Setting   137
From Stockholm to Rio: Environmental Paradigms and International
   Agendas   141
The Clinton Administration and the Greening of GATT and NAFTA   143
Conclusions   146

7 Acid Rain   148

Getting and Staying on the Agenda   149
Bilateral Stalemate and Reagan-Style Hardball Politics   155
Domestic Breakthrough   158
U.S. Action at Last   163
Conclusions   166

8 The Parks Service and the South Moresby Decision   168

The Parks Service as Reluctant Environmental Recruit   169
Protection versus Use, and Old Parks versus New   171
Getting South Moresby on the Agenda: From Local to National
   Politics   175
From Political Theatre to Environmental Bargaining   180
Planning and Pouncing   186
Conclusions   188

9 Environmental Assessment and the Quest for Legislation   190

Early Caution and the Avoidance of U.S. Excess   191
Public Involvement and the 1977 Reforms   197
The Guidelines Order and Its Aftermath   202
The *Rafferty–Alameda* and *Oldman River* Court Cases and Bill C-13   208
Conclusions   210

**10 Legislative Catch-up: From Fish to Toxics to the Canadian
   Environmental Protection Act**   211

Traditional Regulation and Early Agenda Choices   212
Toxics and the Environmental Contaminants Act   216
Fish Wars: The 1977 Fisheries Act Amendments   218
The 'Toxic Blob' and the CEPA Process   221
CEPA and Equivalency Agreements: The Rough Road to the New
   Millennium   225
Conclusions   228

11 Conclusions   230

Federal Environmental-Agenda Setting   231
Environmental Successes and Failures   232
Political Parties, Key Institutions, and Future Decision Making   233
Sustainable Development as a Latent Policy Paradigm   235

Greening, Economic Competitiveness, and Business Power   237
Communicating a Green Public Philosophy and Canadian Unity   239
The Double-Dynamic Framework and Environmental-Policy
    Analysis   240

NOTES   249

SELECTED BIBLIOGRAPHY   271

INDEX   285

TABLES

1 The Legislative Base of the DOE   21
2 Ministers of the Environment   39
3 Legislative Powers of Parliament and Provincial Legislatures
    Relevant to the Management of Environmental Affairs   84
4 Key Events in Acid-Rain Policy   150
5 Key Events in the South Morseby Decision   176
6 Components of the Canadian Environmental Protection Act   224

Preface

The research for this book was carried out primarily during the period from 1989 to 1992 and is based on several sources of information. First, the federal Department of the Environment (DOE) cooperated fully in giving the authors access to internal papers and archives. Although access did not extend to official Cabinet papers, many draft Cabinet documents were read, as were confidential departmental memoranda between ministers and senior officials.

A second vital information source for this analysis was more than 100 interviews carried out by the authors with ministers, officials, and experts in the DOE; in other federal and provincial agencies; and in business organizations, environmental groups, and international institutions. These interviews were carried out on the undertaking that there would be no direct attribution or quotation. The purpose of the interviews, as a complement to other research sources, was both to help reconstruct events and to obtain frank assessments of policy relationships and decisions. We are especially indebted to these many individuals who generously gave their time. Some of them also read all or portions of the manuscript.

The book also builds on a wide array of Canadian and comparative environmental and policy literature as well as published governmental and private-sector reports and studies, as cited in the notes and in the selected bibliography.

While this account of over two decades of Canadian federal environmental policy, politics, and organization is comprehensive, it obviously cannot claim to cover all aspects of the Canadian environmental story. Not only is there some inevitable arbitrariness in the selection of federal decisions for the policy case-studies we present, but there is also only limited reference made to the numerous purely provincial and local government policies in the environmental sphere. None the less, we have tried to find ways to ensure that this book will enable Canadians to better understand the wide range of political and policy problems that confront environmental-policy makers and citizens in this most vital of policy areas.

Special thanks are owed to many long-time employees of the DOE, for whom the department's history is also the heart of their own working lives. They have been remarkably candid about the shortcomings of the department and of Canada's environmental record. But they are also deeply committed to the need for environmental progress and proud of the successes that have been achieved. Above all, they know how long and arduous the remaining environmental journey is for Canadians, and for the planet.

Our indebtedness is especially great to Ken Clark, who helped immeasurably in the organization of the archives and files and in conveying his own experienced and perceptive views of environmental issues both in interviews and in written comments on early drafts. Bob Slater has also been a source of great overall support in encouraging us to undertake the work. Thanks are also due to Glen Toner, Mike Whittington, Katherine Graham, Richard Van Loon, Ken Ogilvie, and two anonymous readers chosen by the University of Toronto Press, whose constructive critical comments on all or parts of the manuscript have greatly helped to improve the final product. None of the above bears responsibility for any remaining inadequacies in the book.

Last, but certainly not least, we gratefully acknowledge the financial support given directly or indirectly to our research by the Social Sciences and Humanities Research Council of Canada, Carleton University, Environment Canada, the C.D. Howe Institute, and the Canadian Environmental Advisory Council.

Abbreviations

ADM	assistant deputy minister
AECB	Atomic Energy Control Board
AECL	Atomic Energy of Canada Ltd
AES	Atmospheric Environment Service, a branch of the DOE
APCD	Air Pollution Control Directorate
CCIW	Canada Centre for Inland Waters
CCME	Canadian Council of Ministers of the Environment
CCREM	Canadian Council of Resource and Environment Ministers
CEAA	Canadian Environmental Assessment Act
CEAC	Canadian Environmental Advisory Council
CEN	Canadian Environmental Network
CEPA	Canadian Environmental Protection Act
CFCs	chloro-fluorocarbons
CO_2	carbon dioxide
CWF	Canadian Wildlife Federation
DEA	Department of External Affairs
DIAND	Department of Indian Affairs and Northern Development
DOE	Department of the Environment
DREE	Department of Regional Economic Expansion
DRIE	Department of Regional Industrial Expansion
EAP	Environmental Assessment Panel
EARP	Environmental Assessment and Review Process

ECA	Environmental Contaminants Act
ECE	Economic Commission for Europe
ECP	Environmental Conservation and Protection, a branch of the DOE
EMR	Department of Energy, Mines and Resources
ENGO	environmental non-governmental organization
EPA	U.S. Environmental Protection Agency
EPS	Environmental Protection Service, a branch of the DOE
ERDAS	Economic Regional Development Agreements
FEARO	Federal Environmental Assessment and Review Office
FPACAQ	Federal-Provincial Advisory Committee on Air Quality
G-7	seven leading Western economies/countries
GATT	General Agreement on Tariffs and Trade
GNP	gross national product
Green Plan	major 1990 federal environmental-policy initiative
GST	Goods and Services Tax
IJC	International Joint Commission
IMO	International Maritime Organization
ISTC	Industry, Science and Technology Canada
ITC	Department of Industry, Trade and Commerce
IUCN	International Union for Conservation of Nature and Natural Resources
LRTAP	Long-Range Transport of Airborne Pollutants
Montebello	Château Montebello, Quebec conference facility and resort
NACE	North American Commission on the Environment
NAFTA	North American Free Trade Agreement
NEP	National Energy Program of 1980
NEPA	U.S. National Energy Policy Act
NOx	oxides of nitrogen
OECD	Organization for Economic Cooperation and Development
Parks Canada	a branch of the DOE, also called the Parks Service
PCBs	polychlorinated biphenyls

PCO	Privy Council Office
PMO	Prime Minister's Office
PPMs	process and production methods
R&D	research and development
RDGs	regional directors general
RIAS	Regulatory Impact Assessment Study
Rio	June 1992 Rio Earth Summit
SO$_2$	sulphur dioxide
Stockholm	1972 U.N. Conference on Human Settlements
UNEP	U.N. Environmental Program
UNESCO	U.N. Educational, Scientific and Cultural Organization
VOCs	volatile organic compounds
WMO	World Meteorological Organization

THE GREENING OF CANADA:
Federal Institutions and Decisions

Introduction

Environmental realities and green political and policy issues have unmistakably moved to the forefront on the Canadian and world policy agendas of the 1990s. What is less clear in the Canadian context is exactly how we got to where we are now in terms of the growing interdependence of environmental and economic issues. Accordingly, this book has a threefold purpose. First, it provides an analysis of the broad political-institutional evolution of federal environmental policy and decision making over the past two decades, beginning in 1970, when the Trudeau Liberal government established the Department of the Environment (DOE), and extending through to the end of the Mulroney Conservative government, including the tabling of the federal Green Plan in 1990 and the holding of the Rio Earth Summit in June 1992. Second, it seeks to provide an enhanced appreciation of the nature of federal environmental policy by linking key institutional relations in environmental policy with the underlying biophysical realities and scientific uncertainties of environmental-policy making. Third, it seeks to add to existing literature a concrete evaluation of the strengths and weaknesses of the federal Department of the Environment, Canada's lead policy department on environmental matters.

In many respects, the book focuses on the trials and achievements of the DOE. But it does far more than that. It also discusses the politicians who tried to advance environmental issues as well as those who seemed only to bide their time. It is about scientists and environmental advocates who have tenaciously committed themselves to making environmental progress. It is about the exercise of economic and political power among ministers, provinces, and countries, and between business and government. And it is the story of numerous concrete decisions taken over more than two decades to address complicated ecological realities.

Even twenty years later, when the Department of the Environment can rightly point to many environmental gains, nature is still broadening the field of battle. Instead of merely bridging rivers in local settings, the environmental challenge now spans oceans and continents. Environmental issues have expanded from local and regional problems of air and water pollution to complex, interactive, and persistent problems that threaten the planet as we know it. Moreover, at the same time that environmental problems became more complex and costly, the political relationships among the main institutions involved in the making of environmental policy (government organizations and non-governmental interests) became more difficult to resolve and manage on the road to environmental-policy action. Even when an environmental issue was publicly defined as a problem, environmental-policy solutions were always hotly contested, or even actively resisted, by vested interests both within and external to government. The resolution of these disputes and power struggles took up valuable time for environmental policy while the biophysical properties of the environmental problems changed beneath the feet of policy makers. As a result, environmental policy never seemed to catch up with the environmental problems; reach always exceeded grasp.

To set the context for the analysis, we introduce here three elements of the journey to be taken. First, we set out the key questions the book seeks to address. Next, we introduce a framework for analysis. And finally, we offer an initial portrait of the key stages of environmental policy since the early 1970s, cast in terms of the rise and fall and rise of the DOE as a policy and operational agency.

Key Questions and Issues

The analysis focuses on six basic questions and issues. First, we probe how the Canadian environmental agenda has been set over the past two decades. This involves an examination of partisan and overall political priorities of the Liberal governments of Pierre Trudeau and those of the Mulroney Conservatives. But it also necessitates a close look at the influence of the scientific community and the environmental and business lobby in priority setting.

Second, we seek to identify, in part through detailed decision case-studies, where environmental progress has been greatest and where failure has been most evident in the Canadian environmental record. This means looking at an array of decisions that encompasses the regulatory, expenditure, and exhortative instruments of environmental policy, and their interconnections, across the more-than-twenty-year period. And it means taking into account both the national and the international dimensions of those decisions, and of greening as a whole.

The third question centres on how much of the success or failure of environmental policy can be attributed, not just to the DOE, but also to key political and institutional players, such as political parties, other government departments, the provinces, business and environmental groups, and international institutions. This involves as well an examination of how these institutions interact in decision making for a complex policy sector.

Fourth, we inquire into the federal government's capacity to advance the concept of sustainable development from its current status as a latent policy paradigm to a more entrenched one, central to all future decision making. The prestigious international Brundtland Commission defined sustainable development as 'development that meets the needs of the present without compromising the ability of future generations to meet their own needs.'[1] Endorsed in principle by the G-7 political leaders at their Toronto Summit in 1988, the concept of sustainable development moved even farther towards the political centre stage at the massive Rio Earth Summit held in June 1992. The concept of sustainable development, in contrast to earlier approaches that focused on cleaning up environmental pollution after it has occurred, inherently raises issues of what kinds of reformed social-decision processes are needed to prevent environmental degradation. This line of inquiry in turn essentially involves an examination of whether environmental governance can be seen even remotely to be the task of one organization as opposed to the collective task of all organizations and individuals. Sustainable development is a latent paradigm in the sense that it is not a fully developed concept, nor does it command majority support among decision makers used to the primacy of the dominant and more established liberal economic paradigm.

Next, and more speculatively, we probe what the keys to environmental progress might be in an age of increased international economic competitiveness and business power. This question involves an understanding of the power of business in past and existing environmental regulatory processes, but it necessarily also relates to the kinds of balances that must be struck in the future between traditional regulation and the use of economic instruments such as taxation, spending, and tradable pollution permits. An examination of this issue in turn involves an appreciation of how much more complex business views of the environment have become in the context of both domestic and international trade.

Finally, we look briefly at the issues of communicating a more integrated green philosophy in an age of mass-media politics and in relation to the potential for environmental issues to become a source of Canadian national unity.

The Double Dynamic

To guide the analysis, we employ a framework which we call the 'double dynamic' of environmental policy. We argue that, in essence, the more-than-twenty-year evolution of Canadian environmental policy involves a continuous tension between two dynamics: first, a time-consuming search – and, indeed, often a long series of power struggles among government organizations and interest groups – to resolve what should actually be done about environmental problems and policy; and, second, the ever-changing ecological and biophysical realm, which is characterized increasingly by unpredictability, scientific uncertainty, and stark spatial realities as the environmental problems of humankind, animals, plants, land, air, and water blend together with great speed and complex degrees of interdependence.

In this section, we introduce the framework. In the next section, we relate our approach to other Canadian and comparative literature on the political and institutional aspects of environmental policy and to the study of public policy. The first dynamic is examined in the first six chapters of the book, where we analyse the political relationships that have evolved among the main environmental-policy institutions and interests, both governmental and non-governmental. In order to promote environmental policy, the DOE had to attempt to resolve and manage a diversity of complex power relationships with: environment ministers; other federal government departments in the Ottawa system; provincial governments; business interests, environmental groups, and the public; and international environmental institutions and other countries.

A further key to understanding the problems that the DOE encountered in managing these political-institutional relationships is to visualize environmental policy within the government as a policy sector (see below). This means that we must look closely at quite specific organizational resources that the DOE did or did not have at its disposal to influence the course of events. The organizational resources discussed at various points throughout this book include: political and bureaucratic leadership; a coherent set of workable policy ideas which can help unify an organization; money and financial resources; legal jurisdiction and statutory capacity; and scientific and technical knowledge.

But even if these complex institutional relationships are resolved, and organizational resources possessed and properly managed, there is still the second dynamic, of biophysical features, scientific uncertainty, and spatial realities, to contend with. There is, in short, the enormous task of coming to grips with the unpredictable and subtle interdependencies of ecosystems and the specifics of key environmental problems such as acid rain, global warming, and the ozone layer, and what particular types of policies are needed to address them. The second dynamic also involves scientific uncertainty, which usually is then connected to problems of how to resolve issues of scientific controversy in public-decision processes. Finally, the second dynamic embraces inherent spatial realities, since pollution occurs in dozens, hundreds, or thousands of physical sites, depending on the precise environmental hazards involved.

Thus, several chapters draw out the problem of the second dynamic. This is done most directly in chapters 7, 8, 9, and 10, which examine four case-studies of key decisions: acid rain; parks policy, and the establishment of South Moresby Park; the search for environmental-assessment legislation; and the forging of the Canadian Environmental Protection Act (CEPA). These decisions span some of the main mandate areas of the DOE and stretch over the more-than-twenty-year time span covered by the book. They have been selected for closer analysis from many possible case-studies because they are exemplars of the struggles inherently involved in environmental-policy making. Acid-rain policy involved complex technical and political relations both with the United States and within both countries. Parks policy embraced highly localized situations that easily evoke national concerns about the balances needed between the use of natural resources and the conservation of Canada's vast ecological spaces. The battle to entrench prior environmental assessment of projects in the decision-making process goes to the heart of the clean-up versus preventative approaches in everyday decision making within the federal government and in business. The DOE's efforts to build an appropriate statutory framework and capacity, eventually through the Canadian Environmental Protection Act, indicates not only scientific and technical dilemmas but also intense federal-provincial relations and key features of the legal aspects of environmental policy.

In addition to these major case-studies, at least ten other policy decisions are examined in a briefer and more illustrative manner in the first six chapters. These include the 1990 Green Plan, Great Lakes pollution, industry regulation in the pulp-and-paper and mining sectors, cuts made to the environmental budget, global warming, the Montreal Protocol and ozone-layer issues, and the Berger Commission of the 1970s.

It could be argued that all departments charged with responsibility for a policy field must confront their own version of a double dynamic and that, with respect to the second dynamic, each faces practical biophysical limits. There is some truth to this argument, especially for natural resource departments such as those responsible for agriculture, the fisheries, forestry, or energy.[2] But it is not difficult to argue that none faces this second dynamic to the extent, scale, or scope inherent in the mandate of an environment department.

Indeed, the very notion of an ecosystem suggests an intricate interdependence and sense of limits. By 'limits,' we mean the continuous process involving the biosphere's flow of energy and its recycling of matter. Ecosystems are in part self-regulating communities in which living members (plants, animals, humankind) and non-living members (minerals, air, and water) interact. The natural environment or biosphere is not merely a part of the economic system. It stands above it and around it because, as Michael Jacobs stresses, it inherently provides at least three services for the global economy.[3] First, it provides humankind with resources (renewable, non-renewable, and continuing). Second, it assimilates our waste products, including not only our unwanted by-products of production but also our wanted products when they too become waste. And third, the natural environment supplies critical environmental services, including amenities for aesthetic and recreational enjoyment and, literally, for life-support.[4]

But even this notion of ecosystems is insufficient to convey the extent to which the second dynamic is endemic to the formation of environmental policy. Environmental issues are also related quite literally to the laws of thermodynamics.[5] The first law of thermodynamics states simply that matter and energy cannot be destroyed or created. The economic system merely transforms one form of energy or matter into another. The second law of thermodynamics (often called the entropy law) states that entropy or disorderliness always increases the more that energy and matter are transformed into wastes. In short, such wastes are more dispersed and less useful than resources. In the final analysis, resources and wastes are quantifiably the same entities, except that the latter have a lower entropic value.

Ecosystem uncertainties are also inextricably bound up with the presence of scientific and technological uncertainty and controversy, and with the role of science in public-policy formation and in regulation.[6] Basic issues of cause and effect; risk assessment; hypothetical occurrences; and how complex issues of science and technology, including the limits of science, are dealt with in policy formation are more persistently prevalent in environmental matters than in other policy fields. The case-study chapters bring out these aspects of the second dynamic, although they are also inherent in our discussion in other chapters of smaller decisions and of the scientific capacity of the DOE.

The key historic conundrum for Canada and for other countries is that even if the key institutions can get their act together, it is not at all clear that nature will cooperate. Indeed, the history of Canadian environmental policy and organization as a whole suggests that the gap between the two dynamics may be widening not narrowing. In short, there is a tension between the two dynamics. It is not simply that the two dynamics are occurring simultaneously. Rather, it is that, in environmental matters, no amount of coordinated political power may be enough. Nor can it be said that the handling of even the first dynamic has been very exemplary.

The Political-Institutional Analysis of Environmental Policy

The book both builds upon and extends the political-institutional aspects of Canadian environmental-policy analysis. In a general sense it clearly builds on key books such as Robert Boardman's collection of essays that embraces key institutional players such as political parties, interest groups, federalism, international relations, and aspects of internal federal decision processes.[7] It also builds on Paehlke and Torgerson's focus on environmental policy and the inherent nature of the administrative state, Doug Macdonald's examination of key features of environmental politics, as well as key earlier work on the DOE by scholars such as Michael Whittington and O.P. Dwividi.[8] A strong political-institutional focus is also found in the work of Ted Schrecker and Kernaghan Webb, who have analysed key aspects of the political economy of environmental regulation in Canada, including key realities about the ability of business interests to resist regulation.[9] George Hoberg's analyses of several aspects of Canadian environmental policy, including its U.S.

influences, is also a vital part of the political-institutional foundation on which this book is constructed.[10] And finally, there is the important recent work by Glen Toner, Jeremy Wilson, and and Filyk and Coté on environmental groups and advisory and stakeholder consultation processes.[11]

In substantive terms, the book extends this base of Canadian political-institutional literature on environmental policy in three ways. First, it more systematically covers key federal institutional relationships and decisions over the two decades since the early 1970s and brings to light sources of research not previously available or examined. Second, it more comprehensively examines, within the context of the federal government, the actual nature of priority setting, the influence of ministers, the structural imperatives of the DOE and the debates within it from the outset, the dynamics of pressure from other ministers and their departments (and the latter's interest-group clientele), and the combined effects of international and federal-provincial relations. Third, the book brings to bear a range of, rather than merely one or two, actual decisions that span the environmental-policy spectrum across the two decades, and that also deal with regulatory, expenditure, and symbolic aspects of environmental governance in combination rather than in isolation.

The conceptual basis of the book similarly both builds upon existing concepts and expands on them. For example, recent Canadian and comparative policy-analysis literature by Pross, Coleman and Skogstad, and Atkinson has profitably employed the concept of policy communities when dealing with any policy field.[12] A policy-communities approach addresses the need to examine policy by mapping a community of policy players and their values. Such a community goes beyond merely a department and its core interest groups. A policy community embraces as well related departments, knowledge networks, and international entities. The notion of policy communities emerged, in a sense, from efforts to be more sophisticated about exactly how society-centred interests interacted both outside the state and, to some extent, within the structure of the state as policy was made and brokered.[13]

To this conceptual base, this book adds a concept which starts from inside the state and, in a sense, reaches outward. In chapters 1 to 6 in particular, where key institutional relations are examined, we show the need to conceptualize a *policy sector* and whether or not lead policy bureaucracies such as the Department of the Environment possess the specific array of organizational resources needed to function and succeed politically and substantively.[14]

A policy sector is a matrix or cluster of government organizations which regularly interact and compete in an effort to defend or promote their policy interests, and which gain access to the policy sector on the basis of their control over organizational resources that are critical for policy-sector activities. These government organizations are influenced by, and often closely associated with, non-governmental interests which may also control organizational resources that are critical for the policy sector.

Organizations which are active within a policy sector are always, to one degree or another, dependent upon other organizations within the sector – in our case, the environmental-policy sector. Since no one organization can ever control all the resources necessary to achieve a policy objective, concrete resource dependencies are at the root of power relationships within a sector.

The concrete resources are precisely those mentioned above in the introduction of the framework, but which often are underemphasized or not incorporated at all in a policy-communities perspective or, indeed, in other policy approaches that deal with political institutions. These resources are: a workable set of policy ideas; political and organizational leadership; money and financial resources; legal jurisdiction and statutory capacity; and scientific and technical knowledge. The analytical difficulty with these policy-sector attributes is that one must deal with them in each of the institutional arenas (the DOE itself, other federal departments, and federal-provincial and international realms). Accordingly, attributes such as a scientific and technical knowledge are dealt with here in several chapters and decisions. The need to have workable policy ideas is found not only in the early debates about the founding of the DOE but also, in different combinations, in how parks policy is formulated, and in both general and site-specific regulatory regimes. Legal jurisdiction and statutory capacities must similarly be examined not only in obvious areas such as the case-studies on the CEPA and environmental assessment but also in their interlocking effects and contradictions in the DOE mandate as a whole.

While this book builds a policy-sector institutional element into the overall framework, a policy-sector concept is not in itself sufficient for the full analysis of environmental policy. For example, key aspects of the larger policy community (business, environmental groups, knowledge constituencies) are also vital and are examined. So also are the analytical insights of economists, especially in the realm of environmental regulation and efficiency in the use of policy instruments.[15]

But the latter perspectives are also insufficient. Hence the need for the second half of the double-dynamic framework, as set out above. Political and institutional approaches to public-policy analysis must also deal with biophysical realms, scientific uncertainty, and spatial realities, and with the temporal stretch of specific and increasingly complex decisions.[16] It is not that other authors referred to above do not recognize this feature. But it is the case that few if any follow through to treat it as a systematic part of any full political and policy analysis of environmental matters. The second part of the framework cannot be treated as mere 'context.' Common sense suggests that it must be woven into the fabric of the frameworks used to understand how institutions deal with the challenge of greening.

The Rise and Fall and Rise of the DOE

The final introductory task is to provide an initial sense of the key stages of Canadian environmental policy cast roughly in terms of the rise and fall and rise of the DOE. The history of the DOE and of the federal environmental effort since the early 1970s can usefully be divided into three periods – 1971 to 1975, 1975 to 1986, and 1986 to the early 1990s – during which the department rose, then fell, and then rose again in public profile, sense of achievement, and political interest. These periods are introduced briefly as a basic, though rough, road-map of the journey followed throughout this book.

The period from the department's inception in 1971 to about the middle of the 1970s was one of growth, enthusiasm for ideas, and ministerial leadership and stability. The main environmental legislation was put in place or further consolidated. Environmental-assessment processes to review projects were established, and considerable international leadership by Canada was in evidence. Aspects of air and water pollution were the main areas for environmental action, with considerable progress made on urban air quality and somewhat more selective progress in water pollution, good in Great Lakes water quality and in phosphorous pollutants but very limited in pulp-and-paper pollutants. The character of the problems was basically local and regional, and related to visible clean-ups of disasters or problems. Political realities meant that the emphasis had to be on reactive environmental measures and on creating a basic regulatory infrastructure. This was not the time for preventative or anticipatory action. There was too much catching-up to do to redress the sins of past unthinking industrialization.

Following this initial burst of energy and drive, the DOE kept a low profile for a decade, with only the brave raising their heads above the environmental barricade. The period from 1975 to 1985 was economically driven, first, by two energy crises and several years of high inflation and unemployment, and then by a savage 1981–2 recession and its aftermath. As a result of both intense industrial and federal-provincial pressure, the DOE turned into what can best be described as a holding operation. There was a high rate of ministerial turnover and hence a lack of consistent political leadership. The Fisheries component left the DOE and was replaced with Parks Canada in the DOE 'org chart.' To all intents and purposes, the heavy gun of the Fisheries Act was lost to the DOE. Even had this not occurred, the federal government was not well disposed to take on the provinces. The DOE was told by Cabinet to leave to provincial environment ministries the main front-line regulatory role.

During this period, the DOE made some small steps towards confronting what they knew to be the new difficult nature of environmental regulation. The Environmental Contaminants Act was passed to deal with emerging toxic substances and other new chemicals being brought to the market-place. But too few resources accompanied these initiatives to permit successful implementation. Indeed, the department went through a series of budget-cutting exercises that severely reduced its regulatory, technical, and service capacities. Research into processes of environmental change and recovery was effectively terminated, except in the atmospheric area – a termination that would cost Canada dearly in subsequent decades. Just at the time when it was evident that environmental problems were now national and continental in scope and that the Canadian public was looking for federal leadership, the DOE was seriously devoid of an identifiable national capacity. Its inability to get anywhere with the United States on the high-profile issue of acid rain symbolized its frustration and its institutional sense of being in a deep political rut. While DOE officials spoke often and seriously about wanting to be able to anticipate environmental problems, they knew in their soul of souls that the rest of the Ottawa system was not listening and had other priorities. Ministerial turnover hit musical-chairs proportions during this period, but the music resoundingly stopped during the disastrous tenure of the first Mulroney environment minister, Suzanne Blais-Grenier.

The period that began around 1986 and took the DOE into the 1990 Green Plan marked the department's resurrection, first quite gradual and imperceptible, and then almost convulsive. The quiet phase included a period during which the DOE's now more stable leadership set out on a deliberate course of cultivating a more supportive clientele and constituency among both business and environmental groups. Several consultative exercises brought the main environmental combatants into a new and more mature relationship with one another.

These important changes were given a vital political boost by a series of international incidents that not only made environmental issues unambiguously global but also simultaneously brought them home to national political audiences. The Bhopal chemical disaster in India, the Chernobyl nuclear-reactor meltdown in the Soviet Union, and the unusual and intense heat and droughts of 1988 attributed to global warming brought environmentalism home with an unprecedented impact. These were accompanied by later, somewhat more national and continental issues, such as the PCB fire at St Basile-le-Grand; the huge tire fire in Hagarsville, Ontario; and, last but not least, the *Exxon Valdez* oil spill in Alaska.

These developments and events led to several reassertions of environmental muscle by the DOE. The first came with the passage in 1987 of the Canadian Environmental Protection Act, which greatly strengthened federal jurisdiction. Finally, following the appointment of Lucien Bouchard as environment minister in 1989, the government and the department launched the process that eventually resulted in the tabling of the federal Green Plan in December 1990.[17]

The Green Plan document tabled by Bouchard's successor, Robert de Cotret, sought to convey to Canadians the new underlying realities of environmental policy. Problems were now global, not just local, or even national. Solutions lay in changing the very nature of decision making in all institutions so as to be more anticipatory. A concept of sustainable development, of 'planning for life,' had to be put in place. Thus, the Department of the Environment began the 1990s by launching a truly ambitious initiative.[18] The Green Plan set out a comprehensive series of goals, more than 100 specific initiatives, $3 billion in new federal money, and many defined targets and scheduled action plans. The plan committed the federal government, among other things, to: the regulation of up to forty-four priority toxic substances in five years; the reduction of the generation of waste in Canada by 50 per cent by the year 2000; the development of new processes to ensure that all federal policies are subject to prior environmental assessment; and the practice of applied sustainable development in the forestry, agriculture, and fishery sectors. Less than a year later, the DOE had persuaded the federal Cabinet to have sustain-

able development put in the preamble of new constitutional proposals as one of the stated goals of a renewed Canadian federation. And,
in June 1992, Canada played a key leadership role at the Rio Earth
Summit.

While the pattern of rise and fall and rise is real enough, and is important to the journey we follow in this book, the reader will also quickly be
aware of the need for subtlety and complexity in telling the Canadian
environmental story. After all, three other realities are a vital part of any
political-institutional analysis. First, the book provides an account of the
lives, ambitions, ideas, and dreams of at least some of the men and
women who constituted the Department of the Environment and the
larger environmental lobby. More than most, these key people understand the imperatives of the double dynamic.

Second, it seeks to provide an understanding of how these same people have articulated, defended, and amended the central ideas that have
struggled for supremacy in environmental politics and policy.[19] It was a
big and politically difficult step for some of these ideas to evolve into a
new, or at least latent, paradigm of environmental and economic policy –
namely, the concept of sustainable development.

Third, the Canadian environmental story is one which fundamentally
cannot be told without juxtaposing the Canadian experience, to some
extent at least, with that of other Western liberal democracies, each of
which was, and is, struggling with similar political and economic circumstances.[20] But this comparative element of the analysis must also be
cognizant from the outset of the differences among countries not only in
how political systems function and in how each country's ecological
composition varies, but also in how the lead environmental department
in each country was structured. For example, the Canadian Department
of the Environment was never like the U.S. Environmental Protection
Agency (EPA) in that the latter was primarily a regulatory agency with
no mandate for aspects of resource management.[21] The DOE was a more
complex entity from the beginning and combined regulatory and integrated resource-management goals and approaches.

Accordingly, the obvious first task of this book is to go back to the
roots of the modern Canadian environmental story and to why the
Trudeau government opted in 1970 to establish a department which
more than one press account initially referred to as being Canada's 'czar
for the environment.' It was a department that was also, from its earliest
days, a house divided.

1

The Department of the Environment: A 'House Divided'

The Department of the Environment (DOE) is the focal point for the analysis in this book, but the department has not been, as we will see, the centre of political power on environmental matters. Its place in the analytical centre is vital none the less because the very structure of the department anchors and encapsulates many of the conceptual and practical issues and tensions of environmental policy. The DOE was put together in 1970–1 from a complex mix of 'organizational orphans' drawn from throughout the federal government, and was then basically left alone to sort itself out and bring some coherence and direction to its activities.[1]

The organizational components of the DOE were first brought together under the old Department of Fisheries and Forestry. In 1970, the Environmental Quality Directorate was established, and, in November of that year, Order-in-Council PC 1970–2047 transferred many other environmental components of the federal government to that department. This ensemble did not officially become the federal Department of the Environment until June 1971. The new DOE thus included seven services: the Atmospheric Environment Service; the Environmental Protection Service; the Fisheries Service; the Land, Forest and Wildlife Service; the Water Management Service; the Policy, Planning and Research Service; and the Finance and Administration Service.

Every major government department is to some extent a house divided. Its constituent units seek maximum independence and have a desire to get on with their roles in an unencumbered fashion.[2] Such divisions and diverse roles can often lead to organizational success, *esprit de corps*, and innovation. But, as always, it is a matter of degree. And, it is a matter also of the extent to which an overall sense of direction guides the department as a whole. The history of the DOE overwhelmingly suggests, however, that centrifugal forces were unusually strong. The account in this chapter of the DOE's internal component parts begins to tell us why. Later chapters, which deal with broader institutional relations, will tell us even more.

This chapter examines the creation and organizational structure of the DOE. The focus is on origins and on the functional and structural components whose differing cultures and inherent policy features are central to understanding both the department and environmental policy. With one or two exceptions, the chapter does not deal with ministers or individuals, or with budgets and personnel. These vital matters come later. Rather, the focus is on five developments that shaped the structure and functional make-up of the department, and hence the making of federal environmental policy: first, the debates over the scope and scale of the DOE's mandate that took place during the formation of the department, and were often revisited as the department evolved; second, the diverse, yet often weak legislative base of the department; third, the departure of the Fisheries Service, and later the Forestry Service from the department, and the reluctant joining of the Parks Service to the department; fourth, inherent disputes within the Environmental Conservation and Protection side of the department, between its regulatory and resource-management wings; and fifth, the problems of integrating the information, knowledge, and service components of the department, centred on the Atmospheric Environment Service.

Trudeau and the 'Mandate Debates'

During the DOE's gestation period in 1970–1 there was considerable debate over what the new department's mandate should be. The Throne Speech of October 1970 envisaged an elaborate role and mandate when it stated that Canada would have 'a department to be concerned with the environment and the husbanding of those renewable resources that are part of and dependent upon it, with a mandate for the protection of the biosphere.'[3] Encouraged initially by Prime Minister Trudeau, proposals for a comprehensive DOE mandate emerged from four sources: the Prime Minister's Office (PMO); the Planning Branch of the Treasury Board Secretariat; planning elements of the DOE itself, mainly the Lands Directorate; and environmentalists, in particular a group of newly emerging environmental lawyers.

The PMO initially argued for a mandate that went beyond a narrow environmental-protection mandate. It saw the new department as an ecosystem manager. These ideas were generated under the supervision of Buzz Nixon, an engineer and strong advocate of systems planning in the then quite rationalist-oriented PMO. The new department, it was recommended, should challenge basic economic-planning assumptions and be able to fill the void of the 'missing ecological feedback loop.'[4] By this was meant a process whereby activities of other government departments that could have impacts on 'air, land, water, energy, plants, or animals' would be assessed by the DOE as to their environmental effects. According to this view, the DOE should then provide feedback to Cabinet and other government departments regarding 'goals, policy, strategy formulation,' 'monitoring, data processing and storage, and prediction,' and 'eco-system research.' This feedback, of course, should ideally have some real impact on decision making within Cabinet and the federal government.

Proposals within the Planning Branch of the Treasury Board Secretariat were almost as grand and were equally enamoured with rational planning concepts. Treasury Board ideas suggested that the mandate of the new department should be to 'ensure for Canadians now and in the future, a desirable state of their physical environment.'[5] However, under the leadership of Douglas Hartle, the proposals were more clearly based in economics rather than in the engineering training of Buzz Nixon. It was argued that a broad and powerful department was needed to instil in others the willingness to incorporate environmental costs into their economic-feasibility calculations. If described in the language of the 1990s, the Treasury Board proposal would have approached the definition but not the scope of a 'sustainable development' mandate. Their document stressed that 'considerable side-effects are not being properly counted as costs, and as a result we probably produce and consume more of the products whose production contributes to environmental deterioration than we would if the full effects are taken into account.'[6]

The new department therefore had to be able to act in ways to 'require alteration in the framework of laws, markets, etc., which govern production and consumption processes.'[7] An argument was even made for the DOE to acquire substantial control over legislation which was, and still is, controlled by other departments within the federal government. The DOE was to involve itself in the direct 'regulation of other government departments and agencies regarding effects on the environment.'[8]

Within the DOE, plans were equally ambitious. One discussion paper trumpeted the new department as one which would 'attack the problems of the deteriorating quality of the Canadian and world environment.'[9] A renewable-resource management role was envisaged, and the DOE was portrayed as the counterpart to other federal departments concerned with monitoring and managing non-renewable resources. As one long-standing senior official of the DOE put it, 'many in the department felt there should be a biological survey to complement the existing century-old geological survey. These types of surveys would then be the analytical base for the management of renewable resources.' This concept was promoted by geographers and resource planners, especially in the DOE Lands Directorate, who wanted to look at the relationship among resources and between the producers and users of resources rather than considering 'one source of pollution and one polluter at a time.' For them it was more sensible to build a 'consolidated department of natural renewable resources.' It was to be a department second to none in its understanding of the natural environment and human impacts upon it.

And last, but far from least, pressure for a more expansive DOE came from environmentalists and, in particular, from newly emerging environmental lawyers. At the 1971 Canadian Bar Association convention, several young environmental lawyers pressed for, and secured the passage of, resolutions which explicitly called for environment departments at the federal and provincial levels that would have the necessary powers 'to review and appraise the various programs' of other departments.[10] Several workshops were also held in 1971 among a loose interdisciplinary group of environmentalists, lawyers, and political activists. This group included individuals such as Greg Morley, Donald Chant, and David Brooks, as well as John Fraser, Lloyd Axworthy, Russ Anthony, and Andy Thompson. Al Davidson and Roy Tinney at the DOE were among the active supporters of this discussion process since it was hoped that it would keep up outside pressure for the more expansive model of an environment department.[11]

As can be deduced from this early 'mandate debate,' the DOE's mandate was very much up in the air even as the department was being pieced together. Prime Minister Trudeau, for his part, was initially enthusiastic about what the environment department could do to define a role and mandate for itself in consort with the central agencies and the other line departments of the federal government.

The prime minister's environmental instincts were influenced by both his general affinity for the outdoors and his early exposure to the work of the Club of Rome. The Club of Rome comprised thirty people from ten countries, including scientists, academics, economists, industrialists, and international and national civil servants, who gathered in the Accademia dei Lencei in Rome to discuss a wide variety of environmental problems. Their famous publication, *Limits to Growth*, had a significant impact in government circles around the world.[12] Trudeau met with representatives of the group at least three times in the months prior to the formation of the DOE.

However, Trudeau's enthusiasm was dampened by the economic and political problems he soon encountered. In response to a question in the House of Commons on 9 October 1970, he began to retreat from his earlier vision of the new DOE. He said that his proposal 'will not result in the creation of a super agency to be responsible for all matters relating to the environment.'[13] His earlier enthusiasm continued to drive comprehensive proposal writing within the central agencies and the DOE, but his actual political support had waned.

The opposition to this larger version of an environmental resource-management mandate came forcefully and overwhelmingly from the ministers and mandarins who headed Ottawa's other main central agencies and resource departments (and the industrial sectors they represented) in the corridors of power in Ottawa. Energy, Agriculture, and Northern Affairs, not to mention other departments, such as Industry, Trade and Commerce, were determined to clip the DOE's wings, and they succeeded.

Trudeau, however, was not the only leader who came to support a scaling-down of plans for the DOE. The department's first team of senior managers also quickly came to support different ideas from those portrayed by federal government planning officials and environmentalists. The department was not to be an 'ecosystem' or 'renewable resource' manager with significant horizontal economic-review powers, but was to be something more akin to an 'end of pipe' service concerned with specific incidents of pollution.

Statutory Capacity

A second key feature of the department has been its varied but often weak statutory capacity. The legal basis of an organization is often given grossly insufficient weight in policy studies. But it is an unavoidable issue in the environmental realm. In this section we simply profile the department's main statutes, highlighting key issues and constraints. Later chapters will explore issues of legal capacity in greater detail. Table

TABLE 1
The Legislative Base of the DOE

Direct Role

Canada Water Act
Canada Wildlife Act
Canadian Environmental Protection Act (CEPA)
Canadian Environmental Week Act
Department of Transport Act (Canals)
Fisheries Act (sections 36–42)
Game Export Act
Heritage Railway Stations Protection Act
Historic Sites and Monuments Act
International Rivers Improvement Act
Migratory Birds Convention Act
Mingan Archipelago National Park Act
National Battlefields at Quebec Act
National Parks Act
National Wildlife Week Act
Weather Modification Information Act

Advisory Role

Arctic Waters Pollution Prevention Act
Canada Shipping Act
Motor Vehicle Safety Act
Navigable Waters Protection Act
Northern Inland Waters Act
Pest Control Act
Transportation of Dangerous Goods Act

Acts Superseded by the CEPA

Clean Air Act
Environmental Contaminants Act (ECA)
Ocean Dumping Control Act

---

1 shows the relevant legislation in the early 1990s, including the acts superseded by the Canadian Environmental Protection Act (CEPA) in 1988. The Fisheries Act, the Canada Water Act, and the Clean Air Act were the most important pieces of legislation for the DOE prior to 1988, but since then the CEPA has become a key legal pillar for the department.

The three dominant pieces of legislation prior to 1988 were *medium-based*. That is, they were concerned with controlling pollution within a specific physical medium – namely, water or air. The Environmental

Contaminants Act (ECA) was a *substance-based* piece of legislation. It was intended to control the use and disposal of particular chemical substances which could enter the environment and travel through ecosystems in a number of different ways. However, as chapter 10 shows, the ECA went almost nowhere in terms of implementation. The CEPA is intended as an advance on this type of legislation. *Product-based* legislation, an area of strong federal jurisdiction, is generally not administered by the DOE, and instead the department has an advisory role to play. Examples of these types of acts include the Motor Vehicles Safety Act, administered by Transport Canada, and the Pest Control Products Act, administered by Agriculture Canada.

The legislative foundations of the department should be further subdivided. The Fisheries Act, as the next section shows, assigned potentially very strong regulatory powers to the federal government. It allowed the DOE, at least in theory, to develop regulations to prevent the dumping of harmful pollutants into water frequented by fish. This act underpinned much of the activities of the Environmental Protection Service. However, from very early on in the history of the DOE, politicians and senior management did not, for various reasons that we explore again in later chapters, deploy this Act to any great extent.

The Canada Water Act and the Clean Air Act, by contrast, were "enabling" rather than "regulatory" pieces of legislation. As we see below, these laws authorized or enabled the Minister of the Environment to reach agreements with the provinces and industry on environmental problems. These acts were based on the spending power of Parliament and the knowledge and information capacities of the administering department. Such enabling acts can be quite effective when budgets are comparatively rich, and/or new knowledge and information can appeal to the provinces or industry under technology transfer agreements. But the decline of these two organizational resources weakens these statutes greatly as later chapters show.

Campaigning to Get Out: Fisheries, Forestry, and the Parks Swap

A third key structural factor of the DOE was centred in the struggle of key units to leave the department and the reluctance of other units to join it. The Fisheries Service, which had its official formation within the DOE on 11 June 1971, would eventually leave the department in April 1979. As compensation for this loss, the DOE was given the Parks Service, transferred from the Department of Indian Affairs and Northern Development under Order-in-Council PC 1979-1617. However, not long after this 'swap' the Forestry Service also left the department. And as recently as the summer of 1993, the Parks Service left the DOE to become part of the Department of Heritage.

With its roots in the old Department of Fisheries and Forestry, the Fisheries Service was a clientele-based organization intent on the management of the fisheries resource for maximum return to fishermen and the Canadian economy. Its inclusion within the DOE stemmed from early interest in a broad renewable-resource management mandate for the new department. Its control of the Fisheries Act was also vital to any ability of the federal government to act on pollution problems, specifically water pollution. On the other hand, scientific, technical, and operational capacities of the federal government regarding the medium of water were predominantly located in the Inland Waters component of the DOE.

However, the main issues to be faced were problems of fish marketing and pollution in fisheries' waters. Fishermen were encountering economically damaging fluctuations in national and international markets for their products. There were too many fishermen and not enough demand for fish. The solution, as the Fisheries Service saw it, was to increase Canada's capacity to manage and conserve the resource off its coastline.

The concern with the economic health of the fisheries industry dominated the activities of the Fisheries Service and led to constant friction within the DOE. Fisheries, as one might expect, continually pushed for a high profile for its clientele in terms of departmental funding. Other organizational components which were more interested in the management of pollution levels originating from multiple users, or in negotiating pollution reduction agreements with industry, were less comfortable with such a pro-regulatory position.[14] Moreover, stressing the needs of the fishery always seemed to imply a loss of scarce resources for other components within the DOE, as we see below.

An organizational culture stressing service to clientele was revealed clearly in an early DOE document.[15] The Fisheries Service outlined several broad policy objectives which were to define their activities. The first priority of the service upon entering the DOE was to 'provide management of the nation's fisheries resources to optimize net benefits to Canada.' This involved the need to:

- Inventory the living aquatic resources in demand by Canadians;
- Maintain a capability to assess demand on fisheries resources, and to regulate their exploitation to provide an optimum sustained economic return to Canada;
- Achieve an equitable return of benefits from anadromous fish of Canadian origin that enter international waters or territorial seas of other nations;
- Achieve Canadian management of living aquatic resources found predominantly adjacent to continental shelves and slopes;
- Increase Canada's share of the fish harvest from adjacent continental shelves and slopes from one third to one half of the total yield.[16]

Other major policy goals included increasing the 'annual yield of indigenous stocks by 50 million pounds in natural waters' and encouraging 'adjustments in the primary fishing industry to achieve the largest gains at the lowest possible tangible and intangible costs.' Fisheries was to remain an operational component in support of the fisheries clientele and the viability of the fisheries industry. Moreover, it was largely an operational and regional or field-based entity, unlike, as we will see later, some other components of the department, which were headquarters-oriented.

Another policy objective that was given priority revealed that the service would 'remove the adverse effects of pollution upon aquatic life.' The specific activities under this objective were said to be:

- Establish contingency plans and capabilities to restore deteriorated habitats and their living aquatic resources;
- Develop a rational basis for the control and beneficial utilization of nutrient discharges to major Canadian Fishing waters, including the Strait of Georgia and the Gulf of St. Lawrence;
- Develop biological indices of water quality adequate for the enforcement of pollution control;
- Establish national effluent standards;
- Clean-up federal facilities;
- Develop a rational approach to solid waste management.[17]

Thus, Fisheries was to be a strong advocate of water-pollution control. As one senior official of the department put it, 'Even before the formation of the department, there were major struggles between the energy department and Fisheries regarding how to deal with pollution. These struggles carried over into the new the DOE.'

Also of vital importance was the fact that the Fisheries Service was in the rare position of being simultaneously pro-development regarding one specific industrial sector as well as a staunch defender of regulatory control of polluting emissions from other industrial sectors. The service intended to be strict in defence of the quality of fish as food and, later, often became disgruntled with slow environmental-protection action in other parts of the DOE. To the Fisheries Service, the strength of the department was the habitat and pollution provisions under the Fisheries Act. It argued that these should be deployed accordingly. To the extent that they were not, serious problems resulted. As one senior official put it, 'Fish were seen as the "miner's canary" for the human race ... Sick fish meant we were all in trouble.'

It was not long after the formation of the DOE that the Fisheries Service revealed its frustration with what it considered to be its demotion from Fisheries and Forestry to one component within the DOE. By January 1973, the DOE went through its first of many major reorganizations intended to placate this discontent. The department was reorganized to contain two major service areas, one of them being the Fisheries and Marine Service under a senior assistant deputy minister (ADM), Ken Lucas, and the other being Environment Services. Lucas was an aggressive promoter of the interests of the Fisheries Service and was a strong advocate of strict pollution control under the Fisheries Act. After the initial 1973 reorganization, Fisheries was still intent upon establishing greater autonomy from the Environment side of the department. A 1974 DOE paper summed up some of these problems as follows: 'a period of organization and reorganization began marked by strong and, in large part, successful efforts by transferred groups to maintain their identity, and also marked by a high degree of what can best be described as jockeying for position by those among the personnel involved who are ambitious for power.'[18]

Many senior officials thought it had been a serious mistake to integrate a specific clientele-industry sector into an 'abstract concept of the environment.' They thought that Fisheries should be pulled from the DOE because it was a clientele department.

The steady movement towards a Fisheries and Environment split in April 1979 had other major consequences for the department. One of these consequences was the consumption of enormous amounts of the time available to senior management of the department. Managing the organizational disruption was a huge burden. By 1979 the situation was even worse. Several senior managers had to conclude, in retrospect, that they could remember the administrative and management issues better than the substantive issues because they had to spend so much time on them. One was even more blunt when he commented, 'Internally [the] DOE was awful; every issue was a jurisdictional conflict.'

Perhaps even more significant, however, were two additional effects. The first of these was the serious threat to the legislative base of the DOE's Environmental Protection Service. The second was the siphoning-off of budgetary resources from the other components of the department. As stressed above, the Fisheries Act was one of the core pieces of legislation upon which the DOE could act on water-pollution issues. The exodus of Fisheries and Marine in April 1979 left the administration of the pollution provisions of this act open to question and resulted in serious uncertainty in the DOE. This uncertainty remains an important aspect of the post-separation relations between the DOE and Fisheries.

As for the resource impacts, Fisheries resources expanded considerably as a percentage of the total DOE budget in the years just prior to the split in 1979. This increase came at a time when the department as a whole was under budget-reduction pressures. The growth of the Fisheries Service, therefore, needed to come eventually out of the budgetary hides of the other DOE components.

The transfer of the Parks Service to the DOE in 1979, right after the Fisheries and Marine Service left, was intended to compensate the department for the depletion of resources. But this in itself created·further organizational problems. The Parks Service was an independent-minded organization which, as chapter 8 shows further, was also heavily operational, with large amounts of resources committed to capital expenditures, many of these presenting prime targets for other DOE senior managers looking for scarce resources. It came into the organization with a great deal of suspicion about whether it was simply going to be a cash cow for a department which saw itself as badly underfunded.

The fears of Parks Service management were not unfounded throughout the first half of the 1980s. Senior DOE management was trying to use Parks Service resources to fill in the many resource shortfalls it was experiencing throughout the organization. This led to considerable internal disruption and difficulty in integrating the Parks Service into the department. Thus, the Fisheries and Marine split contributed to DOE organizational difficulties well beyond its leaving the department.

The extent of the internal struggle between the Parks Service and senior DOE management is revealed clearly in a 1985 memorandum from the ADM Parks, Al Davidson, to the deputy minister of the DOE, Genevieve Ste Marie.[19] At that time, Parks had been in the DOE for almost six years and was continuing to argue strongly against departmental efforts to increase depleted resources through transfers from the service. It was being asked to provide another eighteen person-years (PYs) to the Corporate Policy section.

The ADM reminded the deputy minister of the many sacrifices made by the Parks Service since its arrival at the DOE. He noted that Parks had already given up fifty-eight PYs and related monies to priority needs elsewhere in the department, and that significant expansions in new sites and facilities, supported by politicians and senior management because they were politically attractive, required that resources be 'found' internally 'to the tune of 300 person-years.' This was at the same time that 'Treasury Board imposed reductions of some 380 person-years and 23 million dollars on Parks,' and a budgetary review had already shown the service to be in a 'shortfall of 380 personnel and about $300 million in our maintenance program.' He then noted the effect this was having on the service:

It is demoralizing to Parks Officers to continually attempt to do more with less, to scramble to meet needs, to tighten our belts to scrape up a few more $ and PYs to meet new commitments, and then to find that monies in Parks estimates for planned and approved projects must be transferred elsewhere with little consideration of those Parks needs. Indeed, it can no longer be satisfactorily explained or accepted. We are being taken advantage of for being good managers.[20]

By that time the internal organizational frustration was so high that Davidson concluded:

If it is decided to escape T.B. [Treasury Board] authority by a hidden process to 'transfer' by overutilizing PYs in Corporate and underutilizing in Parks, it should be remembered that T.B. and legislative authority are needed for salary money transfers. If in the final analysis, some further cuts in Parks are decided on I will be forced to make these explicit in service to the public terms in order that the relative value of the projects proposed to be funded and cut, can be judged.[21]

This rather overt threat by Davidson was not one to be taken lightly. Parks was a very popular area of activity for the government and for individual ministers. It had a relatively large and vocal clientele base, including individual members of Parliament who had parks within or adjacent to their constituencies, small tourist-business operators, environmentalists, and aboriginal peoples.

The Fisheries split and the Parks swap were not the only organizational problems faced by the department. The uneasy fit between resource management and environmental conservation and protection expanded into relations with the Forestry Service as well. At the same time that Fisheries and Marine was leaving the department in April 1979, the Forestry Service was reorganized into a full service, with its own assistant deputy minister. The Cabinet also created a Forest Sector Strategy Committee to coordinate the forestry-related policies and programs of the federal government. It had been concluded that leaving the Forestry Service within the Environmental Management Service meant an inadequate profile relative to the importance of forestry and related industries to the Canadian economy.[22]

A friction between the Forestry Service and the Environmental Protection component of the department had existed from very early on in the department's history.[23] Like the Fisheries Service, although not to the same extent because it was a smaller organization, the Forestry Service had felt demoted within one large environment department. It was also concerned with economic-development issues, although, because forestry is in provincial jurisdiction, it was primarily a research organization operating in cooperation with the provinces and the industry.

Shortly after it joined the DOE, the Forestry Service declared its intention to bring to the department major research capacities in support of the provinces and industry.[24] It was going to inventory 'the economic wood supply of the nation' and assist in 'a major reduction in the losses to the economically or otherwise important forest estate caused by insects, disease and fire.' Other research programs included issues of 'reforestation,' 'reducing wood costs,' creating 'a stronger competitive position for Canadian wood products,' developing 'an economically competitive clear wood finish that will retain the beauty of the wood for 15 years exposure to weather,' and increasing 'benefits from the forest resource through development of uses for under-utilized species and currently unused portions of trees.'

The Forestry Service's support of development of the industry created its greatest friction within the department on the question of pesticide spraying. The service was active not only in testing herbicides and pesticides for their effects on the forests and wildlife but also in providing subsidies to the provinces for the spraying itself.[25] This, of course, was long after the publication of Rachael Carson's *Silent Spring*, a book which had had a significant impact on many people within the Environmental Conservation and Protection components of the department. As one senior manager noted, 'the fighting at times became very fierce.' By 1974 a long-range planning document of the Environmental Management Service concluded that 'the [Forestry Service] should terminate its subsidies to the provinces for pesticide spraying.'[26]

However, the friction between the development interests of the Forestry Service and the Environmental Conservation and Protection components of the department was not totally resolved. By September 1984, Order-in-Council PC 1984-3200 had resulted in the transfer of the Forestry Service to Agriculture Canada, where it was assumed that a higher profile could be given to forestry issues within the federal government. Not surprisingly, the effects on the DOE were not benign. Leading up to the separation, the service's proportion of the departmental budget increased considerably. The damage done to other departmental components was much less severe than had occurred with the Fisheries split, but the effect on morale was not insignificant. It also signalled that implementing any broad conception of a 'conserver' model of an environment department was simply not going to happen, at least not at that time and certainly not in a sector, forestry, where many problems were already identified and many more were looming.

The cumulative splits and internal struggles did not make the life of senior departmental management any easier in the 1980s than they had in the 1970s. Senior management in the DOE spent considerable time trying to find ways to help the department without spending any money. Such help required a concerted effort to attempt to consolidate a highly fractured and demoralized organization. A decision was taken to link the Environmental Protection Service with the Environmental Conservation Service (previously the Environmental Management Service, including Forestry) in November 1986 because, as we discuss below, these services were often acting separately or in conflict. The other aspect of the strategy was to change the name of Parks Canada to the Parks Service because, as one senior manager noted, 'no one knew they were part of [the] DOE and Parks Canada never even referred to Environment Canada.' Finally, steps were taken to integrate the Atmospheric Environment Service into the department's other program areas because they 'were essentially a self-contained organization with a head office in Downsview, Ontario.' All of these efforts were designed to create three large, integrated, and relatively equally sized services.

The Environmental Conserver and Protector Roles:
Resource-Use Planning Versus Regulation?

In 1985, an internal DOE management review recommended the reorganization of the Environmental Conservation Service and the Environmental Protection Service. Organizational components within the Conservation Service responsible for Inland Waters, Lands, and Wildlife, had arrived at the new department with established organizational cultures and objectives which did not integrate easily. Further, the two major DOE components responsible for the medium of water, Inland Waters and the Environmental Protection Service, did not agree on the basic objectives regarding this medium. This again created severe management problems. The new combined service, finalized in April 1988, was called Environmental Conservation and Protection (ECP).

The Inland Waters Directorate of the department was and remains the largest component of the DOE's environmental-conservation section. The directorate was formed within the DOE in April 1972 from established components originating within Energy, Mines and Resources (EMR), specifically the Inland Waters Branch and the Water Planning and Operations Branch. Also incorporated within the purview of the directorate was the well-known Canada Centre for Inland Waters, originally formed within EMR in 1967. One senior manager described this centre as 'essentially like a university campus conducting research in support of Canadian water quality objectives such as those set under the International Joint Commission.'

At the time of the formation of the DOE, EMR, as a major feeder organization, had not yet gone through its monumental growth of the late 1970s and early 1980s. It was still essentially a research organization with a relatively small bureaucratic hierarchy. These surroundings had a major impact on the organizational culture and policy objectives of Inland Waters.

At the time of the formation of the department, the management of Inland Waters stated that they planned to bring to the department strong research capacities which would allow them to continue to fulfil their obligations under the Canada Water Act.[27] The objectives of the organization included developing 'fully integrated federal-provincial data collection networks,' and 'information systems providing adequate knowledge and understanding of social and economic implications of current and future water demand and use.' These objectives were to assist in the completion and implementation of 'federal-provincial (and international) water management plans for major rivers, lakes, harbours, estuaries, and coastal waters.'

Central to Inland Waters's ability to achieve 'national water objectives' was its research capacity and the spending power of the federal government. As we have seen, the Canada Water Act is an 'enabling' piece of legislation and it allowed Inland Waters to cooperate 'with the provinces in comprehensive river basin studies' and then subsequently attempt to put integrated resource-use plans into effect. This had been the practice of the component when it was with EMR, and it sought to continue to do the same within the DOE.

Inland Waters also stressed that the department should not try to set regulatory standards for water quality. A 1974 Environmental Management Service planning document stated simply that 'Inland Waters views standards as virtually meaningless because they cannot be enforced.'[28] Instead, the proper course of action was said to be setting general objectives for ambient environmental quality and then planning the various uses of the water resource within those objectives. This required ambitious cooperation with the provinces, the U.S. government, and multiple resource users.

The commitment of Inland Waters to resource-use planning meant that it was highly dependent on its knowledge and spending capacity. Water-use planning involved the federal government in areas of provincial jurisdiction. The provinces could be swayed by good science and shared costs, but their response was always captured by the adage bluntly stated by one senior manager as 'Come with money or don't come at all.' If money or research capacities declined, as they did after 1975, resource-use planning was severely curtailed as to what could be 'enabled' under the Canada Water Act. Further, the efforts of Inland Waters to reach cooperative arrangements with the provinces often did not sit well with the efforts undertaken by the Environmental Protection Service (EPS) to do what Inland Waters advised against, that is, setting regulatory emission standards.

Inland Waters and the EPS did cooperate well on Great Lakes issues and on other issues, but the fact remains that the EPS was essentially established as a standards-setting organization. It was formed in 1972 out of the Air Pollution Control and Public Health Engineering divisions of the Department of Health and Welfare, elements of the Water Sector of EMR, and most significantly, the Water Pollution Control Division and Environmental Quality Directorate of the Department of Fisheries and Forestry. Fisheries elements ended up defining the organizational culture of the EPS, because they had the support of key senior DOE managers. Greater numbers and expertise of its personnel also favoured the Fisheries Service, as did the availability of the core Fisheries Act as one of the only solid pieces of legislation available to the department.

Shortly after the formation of the EPS, managers of the new component stated that they planned to achieve the following objectives: the setting of 'maximum acceptable levels of air quality'; enforcing 'strict control of automobile emissions'; establishing 'stationary source emission standards for all major contaminants that constitute a significant danger to the health of persons'; and ensuring that 'environmental protection systems,' were put in place to protect water quality based on the idea of 'best practicable technology.'[29] Further, the service was going to perform a leadership role within the federal government so as to 'minimize all adverse environmental effects from all federal government works,' and develop 'legislation (e.g. Environmental Contaminants Act), regulations, guidelines, systems and capabilities to evaluate, regulate and control adverse effects of environmentally toxic or hazardous materials.' While the strong regulatory plans of the EPS were later to be modified in practice, the tone was set for internal struggles within the DOE, mainly over the medium of water.

The DOE lacked a unifying role and mandate which could integrate the Lands Directorate and the Wildlife Service as well. The Lands Directorate entered the DOE as perhaps one of the leading organizations on questions of resource-use planning. It administered the Canada Land Inventory, which was the first of its kind in the world. They were also the core component around which the DOE's environmental-impact assessment capacities would be built.

However, because the Lands Directorate was a very small component and was operating in an area of clear provincial jurisdiction, it was not able to develop a legislative base. This meant that it was highly exposed to the Fisheries Service–related budget transfers and other cuts. Further, the inability of the department to build an initiative around the Federal Policy on Land Use, developed by the directorate, meant that the Lands Directorate was prey to other government departments which wanted to assume its role. As one senior official put it, 'Lands was continually moved around the organization and concealed so as to protect it from other departments, like Agriculture, which saw it as their responsibility.'

The Wildlife Service, in contrast, entered the DOE with a very long history behind it, a vocal clientele base; and two pieces of legislation under which to function, the Migratory Birds Convention Act and the

Canada Wildlife Act. However, the Wildlife Service was also a very small organization. When it entered the DOE, Wildlife management had very serious concerns about having its resources weakened. They were also wary of being submerged under a broader departmental agenda which did not share the service's long traditions. Past efforts towards wildlife management had built up a considerable base of ecosystem knowledge. As one senior official outside the Wildlife Service put it, 'They were the only real ecologists we had.'

The concerns of the Wildlife Service were often proved valid. The worst of its experiences occurred under the ministerial tenure of Blais-Grenier, who, as chapter 2 will show, was quite free in offering up the service's budgets in the first round of Conservative government expenditure cuts in 1984–5. The cumulative effect of these types of experiences only furthered the tendency of the service to see itself as distinct but also vulnerable.

The Atmospheric Environment Service:
From Service Provider to Global Researcher

Last, but not least, we come to the final structural component of DOE, the Atmospheric Environment Service. The AES was moved to the DOE from the Department of Transport. Indeed, early in the DOE story there was always left open the prospect that the AES could simply be rolled back into Transport, a move which in the early period appeared more logical. Or, alternatively, it could be given to some other department, such as National Defence or Fisheries. It was originally placed, as the Meteorological Service of Canada, with Fisheries and Marine, in 1874 under a director. In 1936 it was moved as the Meteorological Division to the then new Department of Transport. This move was intended to balance its services between water and air travel. In 1956 it became the Meteorological Branch within Transport. The first act of integrating it into an environmental agenda was to change its name to the Atmospheric Environment Service in the run up to its move to the DOE in 1970.

The AES always appeared to many in the department as a particularly detached entity. This perception was the result of the fact that, geographically, its work was located primarily in its Downsview offices in Toronto and in far-flung field offices. It was not a headquarters entity, as the EPS was. Its detachment also arose out of the fact that the AES was primarily the provider of a service. Particularly in the 1970s, it was not seen by others, nor did it see itself, as being central to environmental policy. This detachment was a concern for DOE management, who had already been battered by the endless internal organizational splits and struggles examined above. One of the key questions which prevailed in relations between the AES and the rest of the DOE was how to integrate the 'Weather Service,' as it was basically labelled, into an environmental-protection agenda.

The challenge that this integration problem presented could be seen through a consideration of what the AES proposed to bring to the DOE at the latter's inception. AES management outlined the service's tradition of being an information source in support of military and civil commercial activities.[30] It said that it was going to continue to provide information, consultation, and advice on weather to a wide range of users from the individual to 'the specialized needs of forestry, agriculture, fishing, construction, and tourism.' The main concerns of the service were to continue to improve on its weather 'prediction systems.'

However, an Atmospheric Research Directorate was established within the AES in 1971 to give a new special emphasis on research dealing with air quality and with the relationships between land, water, and air. And, while in 1972 the AES stressed that it would begin to be concerned with 'air quality' matters, it was not until the major acid-rain, ozone-layer, and global-warming initiatives in the 1980s that the AES actually became a major factor in the DOE's environmental-protection programs. Thus, while the AES had always had extensive technical expertise and was proud of its long traditions in this regard, it had not been known as a research agency as such. This would change as the 1980s evolved.

Another interesting characteristic of the service in contrast to other DOE elements was that its budgetary culture allowed it frequently to profit from environmental accidents or search-and-rescue catastrophes. The AES would immediately rush in, in the wake of such emergencies, to ask for funds, and frequently would get them. Its expenditures were also usually of a fixed-capital variety and therefore were not easily retrenched after the political effects of a catastrophe had died down. The AES budget did, however, shrink as a proportion of DOE budgets through the 1980s, whereas it had enjoyed proportional growth in the 1970s.

Some of the flavour of the AES situation in the DOE can be gleaned from a report given to Deputy Minister Blair Seaborn by Jim Bruce, the head of the AES, in August 1981. Bruce had been its ADM for about a year and had returned to the AES after a thirteen-year absence. He noted that the change from the 'meteorological branch' of 1967 to the 'atmospheric environment' agency he was now heading 'was not just cosmetic.' It signalled, he said, 'a real expansion in our responsibilities to include research and monitoring of air quality and an active participation in the department's roles in environmental assessment.'[31]

Bruce also drew attention to four features of the AES operation. First, he reiterated what almost every observer of the service does, namely, the intense pride and enthusiasm felt by AES employees at being the local 'weatherman' all over Canada, a task involving the handling of more than 10 million requests for information. Second, he expressed amazement and concern at the rapid rate of technological development used in the delivery of weather services. The AES had an excellent technical and scientific pool, but, on the other hand, by 1981 it was functioning with 270 fewer persons than it had in 1977. Third, he stressed that over 70 per cent of the AES budget was for the weather service, but there was little doubt that the air-quality role and the weather service would need more, not less money, and, moreover, it would need a capacity to use these resources more flexibly.[32]

AES programs in the early 1980s were essentially fourfold in nature: the weather services, which were requiring ever-more-specialized kinds of reporting and information, from local lake warnings for boaters to oceanic reports for offshore oil drillers; the atmospheric-research and air-quality program in which the acid rain long-range transport of airborne pollutants (LRTAP) program was central, having just received Cabinet approval in August 1980. On acid rain, the AES saw itself as being 'the lead agency' for the scientific program; the climate program, which was of very modest proportions with only 'one full-time scientist assigned to the CO_2 climate issue'; and the ice-services program, which included early work with Petro-Canada and Dome Petroleum on iceberg activity and surveillance and the publication of the first major Canadian Ice Atlas after ten years of intensive work.

There is little doubt that the AES did play the lead role in the early stages of the acid-rain and larger LRTAP program. As chapter 7 shows, however, as soon as acid rain moved out of its scientific stage and into a more bare knuckle political-negotiations stage, the lead role on acid rain shifted to other parts of the DOE. None the less, the AES earned its spurs on the acid-rain case and prepared the groundwork for its more central role in the as yet unfinished global-warming saga of the late 1980s and early 1990s. It was also the key player in advising the government and the DOE on how to respond to key environmental disasters such as Three-Mile Island, Chernobyl, and Mount Saint Helens.

By the latter part of the 1980s, the AES was describing its role and priorities in ways which were clearly riding the crest of the wave of concern about atmospheric issues and global warming. In a document setting out its strategic plan for the 1990s, the AES now described its role as that of ensuring that 'Canada has adequate information on the past, present, and future conditions of the atmosphere, ice and sea-state to ensure the safety of the public, the security of property, the maintenance and enhancement of atmospheric environmental quality and the greater efficiency of economic activities.'[33]

It related its activities to new governmental priorities for scientific leadership, to concerns over Arctic sovereignty, and to the realities of fiscal restraint. On the first of these, the AES was increasingly disposed to see itself as the leader of Canada's atmospheric and climate research efforts. On fiscal restraint, the AES stressed that, while its weather-service contacts had more than doubled from the early 1980s, to 23 million, these had been achieved with a 10 per cent decline in person-years.

The AES's program structure as the decade ended was the same as at the start of the decade, but the importance of the climate-services program had clearly risen. And the more that the climate and atmospheric research programs gained in importance, the less detached could the AES be from the rest of the department. But this desire to embrace, and be embraced by, the larger integrated environmental agenda of the DOE was a two-edged sword.

On the one hand, as its own strategic plan stressed, the AES knew that 'climate modelling, while improving, still lacks essential scientific knowledge and computing power to be able to integrate information pertaining to oceans, ice and the atmosphere and its interaction with land-surface and biological processes.'[34] But, on the other hand, the political system in the early 1990s was clamouring for what looked increasingly like instant solutions. The AES thus knew that it was no longer just the DOE's cuddly weatherman. It was now on the leading edge of the latest and biggest environmental whirlwind.

Conclusions

The DOE was in many ways a house divided. While some of these divisions flowed from the original debate about the mandate the DOE would be given, many were also related to matters of internal organizational logic and conflict. A varied and often weak statutory capacity also contributed to policy problems. Some of its constituent services, such as the Fisheries Service and the Forestry Service, campaigned to get out, and succeeded in their quest. Others, such as Parks Canada, were reluctant to join the DOE for fear that their resources would be raided. Still others, such as the AES and the Environmental Management Service, were driven by their very characteristics as services to seek their own realms of independence both operationally and philosophically.

While the organizational architecture of the DOE is an essential starting-point for any account of federal environmental-policy making, there is no doubt that explanations of the rise and fall and rise of the DOE also lie elsewhere in the key political relations that it faced in the larger world of environmental politics.

2

Ministers, Mandarins, and the Green Agenda

While our initial portrait of the Department of the Environment (DOE) as a house divided is an essential one, the discussion of the most immediate environmental-policy relationship, that between the DOE's ministers and senior mandarins, reveals a series of efforts to develop a more concerted green agenda. Every federal environment minister since Jack Davis – fifteen more in twenty-two years – has attempted to set his or her agenda. Most of the environment ministers had days when they hoped that they might be environmental generals. But they quickly discovered that they were usually more like platoon leaders, a small band rarely on the offensive, always defending small but precious gains in the environmental manoeuvres, or alas, suffering setbacks.[1]

In addition, there certainly were numerous efforts by the DOE's deputy ministers and other senior mandarins to try to obtain a greater sense of coherence by developing various environmental plans, culminating in the 1990 Green Plan. Thus the nature of federal green-agenda setting is essentially a linked, two-part story. In the first half of this chapter, we analyse it chronologically as essentially a story of the relations between successive ministers and the DOE's deputy ministers. This part is inevitably an incremental political story as the various preferences for action and inaction were worked out in the context of a larger political-economic world. In the second half of the chapter, we turn to the more concerted efforts of the department as a bureaucracy and as a scientific and technical entity to plan more rationally an environmental agenda.

But at the outset, one stark ministerial fact emerges from the simple listing of ministers in table 2. Sixteen ministers in twenty-two years means that, on average, ministers stayed just over sixteen months. But if leadership is quintessentially a political task, then this pattern of short

TABLE 2
Ministers of the Environment

Ministers	Date Position Assumed
Jack Davis	June 1971
Jeanne Sauvé	August 1974
Romeo LeBlanc (Minister of State, Fisheries)	August 1974
Jean Marchand	January 1976
Romeo LeBlanc (as Department of Fisheries and Environment)	September 1976
Len Marchand (Minister of State, Environment)	September 1977
Len Marchand	April 1979
John Fraser	June 1979
John Roberts	March 1980
Charles Caccia	August 1983
Suzanne Blais-Grenier	August 1984
Tom MacMillan	August 1985
Lucien Bouchard	January 1989
Robert de Cotret	September 1990
Jean Charest	April 1991
Pauline Browse (Minister of State, Environment)	April 1991
Pierre Vincent	June 1993

tenure and high turnover suggests that ministers barely had time to lead or to learn. When one adds the fact that many ministers were junior or inexperienced when they held the portfolio, then the probabilities of leadership are further diminished. There were exceptions to these trends, to be sure, as this chapter shows, but the parade of ministers is itself a dilemma of agenda setting.

The turnover among the DOE's deputy ministers was not as high. Since 1971 there have been six deputy ministers. These mandarins contributed more continuity to the leadership picture but always in ways constrained by the ministerial dynamics and conditioned by the managerial style and abilities of the deputy minister.

Jack Davis and the Engineer

In 1970, Jack Davis was a minister in the middle rungs of the Cabinet ladder in terms of experience and clout but, in terms of professional background, a remarkably good choice to launch Canada's new department. As a resource economist and as an already serving minister of Fisheries and Forestry, or 'fish and chips' as it was then often called, he was already concerned about the ultimate resource dilemma, the proverbial 'tragedy of the commons.' In the case of the fishery it was too many fishermen chasing too few fish, and thus Davis began some of the thinking that later led to the establishment of the 200-mile fishing limit around Canada's coasts.

Jack Davis brought with him as his deputy minister Robert Shaw, an engineer by training. Davis fought for the appointment of Shaw over the preferences of Prime Minister Trudeau, who suggested that someone like an ecology professor should get the job. Shaw had gained considerable fame as a vice-president of Expo 67 and for bringing other large projects to fruition. He was a project manager par excellence. And he was relatively new to the Ottawa scene. Davis picked him mainly because he thought he would be good at dealing with the business community. In this regard, Davis's instincts were quite correct. Several task forces were set up with sectors such as pulp and paper and chemicals, areas which Davis knew would have to be the first to be regulated.

But Shaw had little patience for either the details or the paper flow of the Ottawa system. As a result, he was frequently away from the office and was seen by his assistant deputy ministers (ADMs) as a ghostly fleeting presence. The day-to-day running of the department was left to Jean Lupien, whose title was senior ADM but who in fact occupied a hopeless no man's land. On the one hand, he had little authority; on the other hand, he was charged with keeping the warring sections of the department at bay.

None the less, in this setting, Davis played a remarkably skilful role. He was at times both minister and deputy minister rolled into one. His weekly early breakfast meetings would often bring officials together who would not otherwise meet face to face with the minister. He dealt bilaterally with his ADMs without going through his deputy. He had no compunctions about calling on regional officials on the grounds that such people were far more likely to have actually seen the environmental problem or the physical situation involved.

In the international field, Davis, in concert with Shaw, was instrumental in ensuring that Canada played a pivotal role in the 1972 Stockholm Conference. This included engineering the appointment of Canada's Maurice Strong to head up the Stockholm program. Thus, Canada was put quite conspicuously into a leadership position and, as a consequence, expectations were raised, inside and outside Canada, many of which could not later be met.

In these various ways, Davis did establish an initial *'esprit de corps'* in the department. Moreover, in the early 1970s, when federal budgets were flush, Davis had no trouble getting the needed resources. He fought for them himself rather than leaving it to others. It is of course true that the department had its own momentum, arising out of the fact that it was new. Ideas were encouraged, including some pretty far-fetched ones. In the DOE buildings in Hull, one could sense the different culture of the DOE just on the basis of what people wore. In the elevators, neatness and dark suits were often outnumbered by beards and slacks or jeans and sandals. But there is little doubt that leadership and interest came from the top and that Davis's three-year tenure was successful in setting up the fledgling ministry and directing it to its tasks. Such leadership and continuity were not to reappear for another decade.

Blair Seaborn and the Liberal Group of Six

From 1975 to 1982, the leadership of the DOE can truly be said to have resided with Blair Seaborn, the deputy minister for most of this period, rather than with the brigade of mainly inexperienced ministers, six in eight years. They were inexperienced in two senses. They had had only limited ministerial experience before they arrived at the DOE, and most of them stayed for such short periods that their environmental experience could not be nurtured. This is not to suggest that each minister did not do something to advance some part of the agenda as he or she saw it or as it was presented by the department. But it is to say that most ministers, willingly or otherwise, were engaged in an environmental holding operation coincident with the lower place on the government's priority list that environmental issues had. The first evidence for this was again the simple fact that there were six ministers in eight years.

Some of the flavour of this situation can also be found on the first occasion when Jeanne Sauvé, Jack Davis's successor as minister, met Blair Seaborn. Sauvé, later to become governor general of Canada, had been minister for several months and told Seaborn that she thought it would be a unique situation in that, for a few weeks at least, she would know more about the environment than her deputy. It was indeed unique. Neither she nor her successor, Jean Marchand, had the time or the influence to contribute much to the department's sense of direction. Marchand, once dubbed, along with Gérard Pelletier and Pierre

Trudeau, one of the 'three wise men' from Quebec, had been a powerful Liberal minister but was now on the final stages of his political career. He was being parked in the DOE, where he stayed for six months.

The DOE's next minister, Romeo LeBlanc, was a much more skilled and determined minister and gave the department a new moment in the political sun but one, as chapter 1 has already hinted at, that was a mixed blessing. LeBlanc, was a first-class regional minister. As a New Brunswicker, he made the Atlantic fishery his personal bailiwick for much of his political career in Ottawa. He was first appointed in 1974 as Minister of State (Fisheries), reporting to Jeanne Sauvé in the DOE, and then became the DOE's minister himself in September 1976. The Fisheries component of the DOE had squirmed, as chapter 1 showed, under what it regarded as the yoke of the DOE's existence and had constantly pressed for as much independence as possible. It found a willing champion in the engaging New Brunswicker who knew that opinion in Atlantic Canada and on the west coast supported a greater visibility for the fishery in the Ottawa bureaucratic structure.

Blair Seaborn knew that the Fisheries component was his biggest problem when he had come to the DOE eighteen months earlier. At an early deputy ministers' luncheon meeting, at which Trudeau was a special guest, Seaborn asked the prime minister what he wanted to do about the Fisheries component. Trudeau's instruction was that it was to remain a part of the DOE. But LeBlanc pressed relentlessly, and when he became minister, he headed a department whose title he then changed to the Department of Fisheries and the Environment. Thereafter, LeBlanc signed most of his correspondence 'Minister of Fisheries.' His instructions to Seaborn were equally clear. Fisheries issues would be handled by LeBlanc, advised directly by Ken Lucas, the senior ADM for Fisheries and Marine Services. The smaller Fisheries tail was now wagging the larger DOE dog. This state was symbolized further in September 1977 when Len Marchand was named Minister of State (Environment) and simultaneously the Trudeau government announced its intention to establish a separate Department of Fisheries and Oceans.

None the less, LeBlanc did fight for and achieve one significant environmental-policy change, the establishment of a 200-mile fishing limit on Canada's coasts. While the gestation period for this initiative predates LeBlanc, there is little doubt that he steeled the nerve of the Trudeau government in Canada's unilateral decision to take this action, which took effect on 1 January 1977.

Once the short burst of LeBlanc energy had passed, the DOE settled into another series of junior ministers. Len Marchand, the first aboriginal Canadian to be appointed to the Cabinet, became minister in April 1979. Marchand took some understandable and long-overdue interest in the links between Native People and environmental policy. This included his concerns that wildlife policy did not trample on traditional Native hunting and fishing rights. He also helped advance the acid-rain file within the department. But, in general, he was far too junior to have much impact in his short stay at the DOE.

The DOE's optimism soared, however, when the Clark Conservative government's first and only environment minister was announced in 1979. John Fraser was a legitimate environmentalist. A Vancouver law-yer who had fought many environmental cases and who was a member of several environmental groups, Fraser was one of the few DOE minis-ters who actually wanted the job. For him, the chance to be environment minister was a dream come true. Fraser had well-developed ideas about environmental reform and an intensity about carrying them out that was immediately communicated to a department anxious for political lead-ership. During his brief tenure, environmental non-governmental orga-nizations (ENGOs) enjoyed unprecedented access precisely because here was a genuine ENGO member sitting in the minister's chair. Alas, the euphoria ended resoundingly with the sudden and unexpected defeat of the Clark government on a December 1979 budget vote in the House of Commons.

The contrast between the Fraser appointment and the new Liberal government's first appointee could not have been greater. While John Fraser had wanted the job, John Roberts did not. The Toronto MP had been secretary of state in the previous Trudeau government but had actively sought more prestigious positions in the Cabinet. A former for-eign-service officer, Roberts brought with him the reputation of being a smooth Toronto dandy, politically bright and ambitious but also a bit politically lazy. He was a Trudeau disciple interested in some of the same things that had attracted Trudeau to political life. These included consti-tutional issues and a strong interest in getting a more planned response to public-policy problems.

During his tenure at the DOE from 1980 to 1983, Roberts also held the position of Minister of State (Science and Technology). And he was heavily engaged during 1981–2 as one of the main English-Canadian ministers in the constitutional negotiations that then dominated the Trudeau agenda. The new Trudeau government was clearly pursuing an aggressive new agenda, but the environment was not a central feature of it. Indeed, its actions in the National Energy Program of 1980 ensured that resource development would not brook any environmental second thoughts.

Like most of his predecessors, Roberts brought no particular agenda with him. But gradually he did develop one or two points of interest which he sought to advance and move through the system. Because of his interest in planning, he sought to pick up some of the slack he felt had developed after the early Trudeau interest in rational government. He believed that the environment was a logical place to re-energize these thoughts because he was convinced that environmental policy had to move beyond reactive activities and to engage in more systematic thinking. Thus, Roberts had some of the same intellectual urges for change that both early and later Green Plan designers knew would some day have to occur. But there was little will to carry these ideas forward because to do so would ultimately mean taking on more frontally other centres of power.

In the end, however, Roberts did find one key issue that he shares credit for advancing. This was the acid-rain file. After the Reagan Administration took office early in 1981, it was plain that taking action on acid rain and on other bilateral environmental issues was going to be very difficult. Roberts took up the battle and made the acid-rain issue his own. He supported the development and funding of the Acid Rain Coalition and launched a lobby of U.S. congressional and Senate representatives that was one of the first to go well beyond the bounds of typical Canadian quiet diplomacy in Washington.

Roberts's tenure was the last of the group of six ministers who coincided with the period in which Blair Seaborn served as deputy minister. A career diplomat by training, including peace keeping duties in Vietnam, Seaborn was seemingly sent as a peace keeper to the DOE. His own personal style and the ministerial musical chairs he faced left Seaborn little choice but to be a process man. In comparison with the Shaw era, under Seaborn, the DOE's internal executive-decision processes were greatly smoothed out and the more overt internal friction among the branches was reduced considerably. A great deal of the lessening of friction, however, was also simply a natural product of the fact that the Fisheries component was now gone, and that the branches had now hunkered down into their own operations, fearful of budget-cutting raids and other restraint measures increasingly prevalent in the early 1980s. Seaborn found himself constantly preoccupied with a seemingly never-ending array of federal-provincial issues as well as other ministers' perpetual concerns about forthcoming environmental assessments of projects in their regions or policy domains. In each of these situations, he found himself caught in the middle. He had both to advocate the environmental policies he was charged with administering and, at the same time, soothe many a fevered brow with assurances that things would not be as difficult as they looked for various anxious industries, ministers, and provinces.

Seaborn also presided at the time when the nature and severity of the environmental problems with which the DOE had to deal could no longer be easily classified into simple categories such as air versus water or federal versus provincial. In dealing with acid rain, and with many other issues as well, Seaborn's problem was one of fitting increasingly integrated ecological concerns into competing organizational compartments.

Following John Roberts's departure as minister, yet another type of personality became the helmsman of the DOE. Also from a Toronto riding, Charles Caccia – the last of the Trudeau environment ministers – had a year in office that consisted of modest but fairly thoughtful action, which coincided with his own quiet, unassuming approach. Caccia was no political firebrand, but his own professional background in forestry and his determined commitment to environmental-protectionist values earned him considerable respect in the department. At the same time he knew his own limits and also that an election was less than a year away. His approach was basically to move as many issues forward as possible. Indeed, one of his first acts was to hand his officials a list of some twenty-five items on which he wanted action.

Caccia lent his support to the establishment of the SO_2 '30 Per Cent Club,' a group of countries prepared to commit themselves to reducing sulphur-dioxide emissions by 30 per cent. He was the first minister, as we will see in chapter 9, to throw his weight behind the lobby to create South Morseby Park. He gave full support to Canada's participation in the Brundtland Commission and also launched a study, over the strong objections of Eugene Whelan, the minister of agriculture, on Canada's water resources. The resulting Pearse Commission report was, without doubt, the most comprehensive public examination of this significant environmental and resource issue.

Caccia also had very specific views about the role of economic analysis in resource and project assessments. He knew that his officials increasingly had to speak an economic language to get anywhere with the central agencies in Ottawa. But, at the same time, he wanted his officers to challenge things such as the discount rates used in central agency and other departmental analyses, and the way in which costs were underplayed, especially on large projects.

Caccia was therefore a steadying hand in the Trudeau era's last days of environmental stewardship. But, beyond the mid-1970s, the Trudeau Cabinet was not a bastion of environmental progress. Caccia could count on active environmental support from barely three ministers, among them Trudeau himself, Romeo LeBlanc, and Jean-Luc Pépin. The rest were preoccupied in the 1980s with other issues, not the least of which was the survival of the Liberal party after John Turner was chosen leader and after the crushing defeat of the party in the 1984 general election.

Early Conservative Ministers: Nowhere to Go But Up

Suzanne Blais-Grenier will likely forever be the low point among federal environmental ministers. New to both politics and Cabinet office when she was named by Brian Mulroney as his first environment minister, Blais-Grenier adopted an approach that became a total enigma to all who encountered her. The Quebec-based politician was a former civil servant herself, but she seemed to go out of her way to alienate her own officials. Some of this combativeness was undoubtedly picked up on cue from the general approach being taken by a new Mulroney Conservative Cabinet that was extremely suspicious of the bureaucracy after being out of power virtually since 1963. Mulroney had instituted a new chief-of-staff system whereby ministers would be given highly paid senior policy advisers in their own offices as a hoped-for counterweight to the senior bureaucracy. In the early months of the Mulroney tenure, deputy ministers as a whole were being kept at arm's length from the Cabinet and were being monitored carefully by regular meetings of the newly assembled chiefs of staff.

But Blais-Grenier took these cautious instincts well beyond the bounds of sensible behaviour. She and her office created a siege mentality to such an extent that she did not even want to be briefed by her own officials. It did not help either that she did not get along personally with Jacques Gerin, the deputy minister who had succeeded Blair Seaborn in 1982. Blais-Grenier also wanted to show that she was in charge by being a loyal trooper to those in the new government anxious to impose expenditure cuts. This in itself is hardly grounds for complaint in that many ministers had had to cooperate reluctantly with demands from the centre for expenditure cuts.

The political sin of Blais-Grenier was that she embraced the cuts both enthusiastically and clumsily. When the Treasury Board asked for cuts in 1985, the DOE minister was the first to volunteer for the guillotine, without even checking with other ministerial colleagues as to just what proportionate cuts might be being offered elsewhere. Moreover, she offered a 25 per cent cut in the Canadian Wildlife Service without any discussion as to whether this was the best place to cut. From the time of her cuts until she was hounded out of office by a baying parliamentary opposition, the Wildlife cuts became the *bête noire* of the DOE and brought its morale plummeting to rock bottom.

Blais-Grenier also angered the environmental lobby in several other ways. When approached by environmentalists about federal support for the establishment of a national park at South Moresby, her meagre financial offer was made conditional on the ENGOs raising part of the costs of the park from public contributions. The estrangement between Blais-Grenier and the environmental lobby grew so great that her fellow Conservative minister John Fraser, both when he was minister of fisheries and later, when he was Speaker of the House of Commons, had to privately intervene several times to reassure the ENGOs that Blais-Grenier did not represent Conservative environmental thinking.

Accordingly, Blais-Grenier's successor, Tom MacMillan, had nowhere to go but up. A Prince Edward Island MP with little ministerial experience, MacMillan was not the most likely minister to lead environmental policy to the promised land. But when he left in 1988 he had earned the reputation as being one of the DOE's better ministers.

MacMillan knew that the DOE would not necessarily fare very well if it sought to recover its lost budgetary resources. And he properly gauged that he did not have the clout to get them anyway. So, instead, MacMillan focused on things that did not directly cost money by developing a more credible regulatory approach and fostering a more consultative approach with both business and environmental groups. The agenda he followed was based both on his own instincts and on one of the beneficial aspects of the Blais-Grenier débâcle. This was the major review of the DOE that emanated out of the government's larger Nielsen Task Force program reviews.

Among these Nielsen studies, that on the DOE was the only one that focused wholly on one department. The reason why this was deemed necessary was that there was now a clear perception of ineptness at the

DOE, but at the same time those at the centre knew that public interest and concern about the environment were growing. The initial Nielsen study was followed by a second, and the two sets of recommendations were taken by MacMillan to Cabinet. This led, along with other developments such as the Brundtland Commission, to the adoption of a three-pronged DOE approach that helped pave the way for the DOE's run at a formal Green Plan in the late 1980s and early 1990s. In December 1986, Cabinet approved: a federal Environmental Quality Policy Framework, an economy–environment partnership consultation strategy, and a proposal for a draft Environmental Protection Act. By the end of this process, MacMillan had earned his spurs to the extent that he was able to obtain exemptions for the DOE from some of the continuing five-year plan of cuts still being imposed by the Treasury Board.

While MacMillan worked the outside network and restored departmental pride and coherence, the two deputy ministers during this period tended to the internal needs. Jacques Gerin was, to some extent, victimized by his battles with Blais-Grenier, but he was not viewed in the department as being very decisive or a born manager. He did contribute conceptually, however, to the need for the department to develop its outside clientele. Indeed, these were views he had advanced much earlier in the 1980s, when he was the ADM Policy and was instrumental in developing the DOE's first policy on public consultation.

Gerin is acknowledged by many ENGO spokespersons as being the one senior official in the DOE who most consistently sought to assist the ENGOs in whatever way he reasonably could, whether through financial assistance or through access to information. More than most DOE senior officials, he saw the environmental lobby as an ally and was often frustrated by the department's overall failure to regard them as such.

The deputy minister from 1986 until May 1989 was Genevieve Ste Marie. Her instructions from the centre were to integrate the department and improve its managerial capacity. In this task, she was credited as being a considerable improvement over Gerin. And she helped shepherd and sell MacMillan's concepts to the department, and the department's anxieties to MacMillan. Both she and her minister shared the trait of being good listeners but not very imaginative or decisive decision makers.

What constituted good or better management also depended very much on the eye of the beholder. Ste Marie, for example, did not win plaudits from the environmental lobby. She attempted to cut the funding that the DOE had been supplying to the ENGOs, thus confirming in the minds of many ENGO leaders that the DOE had no institutional memory about, or commitment to, its relations with the ENGOs. Every deputy minister, in ENGO eyes, played a new game.

In a fundamental sense, the department also knew that Ste Marie was new to deputy-ministerial duties and hence lacked clout among the senior mandarins of Ottawa. At the same time, however, Ste Marie was experienced enough to know that large amounts of new money were unlikely to fall into the DOE's hands in the foreseeable future. Accordingly, she saw the value of trying to forge major legislative reform such as the CEPA, which seemed costless in the short run in budgetary terms but which might force new resources out of the central coffers in a year or two once it became clear that the legislation, once passed, then had to be actually implemented. This 'statutory change now and money later' strategy did influence MacMillan's overall approach to bringing the department out of its Blais-Grenier–era depression.

Lucien Bouchard and the Green Plan Ministers

The decade of the 1980s ended with the tenure of Lucien Bouchard, who was minister of the environment for just over a year before he left the Mulroney Cabinet over the Meech Lake Accord and formed a new rump bloc of separatist Quebec MPs. A close friend of the prime minister, a former Canadian ambassador in Paris, and a man known for his strong views, the new minister was unique in the annals of the DOE. He strode into the DOE knowing that he had a prime-ministerial mandate to devise a Green Plan and he set about doing so.

To the gusto of the minister was added the considerable bureaucratic muscle of the DOE's new deputy minister, Len Good. An economist, Good also knew the corridors of bureaucratic influence better than any previous DOE deputy. He had been the deputy secretary to the Cabinet Committee on Priorities and Planning and, prior to that, the ADM Policy in the Department of Energy, Mines and Resources.

But before this formidable team could get to work, Prime Minister Mulroney had to be convinced that greening deserved a higher profile. The Mulroney Conservatives had been re-elected with a majority in November 1988, and in the April 1989 Throne Speech, Mulroney promised a major initiative. But the brief Throne Speech paragraphs devoted to the environment contained only a series of specific initiatives to be pursued. There was certainly no reference to a more comprehensive approach. The incumbent environment minister, Tom MacMillan, had been defeated, and the prime minister named Lucien Bouchard to the Environment portfolio. Mulroney also established the first Cabinet committee on the environment and named the environment minister to the powerful inner cabinet (the Cabinet Committee on Priorities and Planning). While Bouchard had been a member of the government and Cabinet only for a short period, his long-standing friendship with the prime minister had already established him as a senior minister.

On becoming minister, Bouchard was extremely frustrated working in an incremental mode and in what was essentially a policy vacuum. At the bureaucratic level, the DOE had very little capacity to develop policy in an Ottawa system-wide context and, moreover, had no economic- or fiscal-policy strength. Indeed, the Corporate Policy Group at the DOE had been without a Director General (Policy) since January 1988. Nevertheless, the ADM Policy, with a small team, crafted a document outlining a federal agenda for the environment for the second mandate. Intellectually, the argument integrated the environment and the economy by linking the mutually reinforcing twin deficits of the budget and environmental degradation, showing the dubious legacy the present generation was leaving for future generations.

This initial effort was seen as conceptually and politically inadequate. As a result, Bouchard wrote a long letter to the prime minister in February 1989, outlining the intellectual and political rationale for a decision by the government to introduce a major and comprehensive environmental policy for Canada. The Department of Finance drafted a response to the DOE document, which showed just how large the gap was between the DOE and officials in other departments. At this point the Ottawa policy system simply did not believe that the environment was going to emerge as a dominant issue for the government. Bouchard realized that he needed more analytical support to deal with the environment–economy issues, and in May he brought in Len Good as his deputy minister. The new DM set to work creating a team of policy experts from around the Ottawa system, drawing heavily on those with experience in major economic departments or the central agencies.

There is little doubt that, in the context of 1989–90, no one other than Bouchard could have got a federal Green Plan off the ground. But his tactics, and the juggernaut secretive approach he used internally also produced tremendous resistance among other ministers and departments. With his abrupt departure in 1990, the Green Plan could easily have fizzled.

It was left to the unexpected and underestimated skills of his successor,

Robert de Cotret, to protect the Green Plan and garner the resources needed to give it credibility. The autumn 1990 period was consumed by a very intense series of financial negotiations with the Department of Finance at both ministerial and bureaucratic levels over the global figure for the plan, and with each department over its share of the money and jurisdiction. As the overall figure moved down to the eventual $3-billion price-tag, each department's programs and share had to be recalculated, with every department suspicious that it was giving up more than anyone else. These negotiations were extraordinarily tough, with extensive conflict present among senior officials. De Cotret had to call in every political IOU he had acquired as president of the Treasury Board to get a deal. It was also during this period that Cabinet altered the structure of the document by placing the big health, clean-up, and spending programs in the first part of the plan and burying the decision-making section in the back. This decision would later come back to haunt the government, when critics charged the plan with lacking a 'vision' of sustainable development and for once again throwing money at a policy problem.

The Green Plan was tabled in the House of Commons on 11 December 1990. Over the previous eighteen months, it emerged through a difficult gestation involving two ministers, and, as we see further in chapters 3, 4, and 5, fierce interdepartmental battles; intense federal-provincial conflicts; and a high-profile, controversial consultation stage.

The last Mulroney-era minister of the environment was Jean Charest, who held the post until his unsuccessful bid for the leadership of the Conservative party in June 1993. As a young engaging minister, Charest was seen by his department as well suited to selling the Green Plan. While Green Plan announcements were somewhat slow in materializing in 1991, they became more numerous in 1992 and earned the government a better environmental report card in public opinion than had been evident in the immediate aftermath of the Green Plan's announcement.[2] Charest also garnered considerable credit for the quite successful Canadian leadership shown at the Rio Earth Summit in June 1992.

Somewhat puzzlingly, however, Charest's leadership bid for the Conservative party seemed to signal the potential decline of the environment as a political issue in 1993. Despite being the environment minister, in his key leadership speech to the Tory Leadership Convention he made not a single reference to green issues. Nor did the eventual winner and Canada's first female prime minister, Kim Campbell.

Suffice it to say in this initial account of ministers and mandarins that the DOE entered the 1990s largely on a political high. We leave to later chapters our discussion of whether this new heady plateau is likely to be sustained in the remainder of the 1990s.

Departmental Planning and Learning

The parade of ministers and deputies, strong, weak, and indifferent, gives only one glimpse, albeit a telling one, of the basic political dynamics of green-agenda setting. Also going on throughout the twenty-year period were numerous formal, concerted efforts by the DOE as an organization to plan its future and to think through both its general mandate and priorities and those of its constituent parts. If ministerial prisms supply a top-down view of priority setting and concerns, these planning exercises complement that view with a bottom-up or organizational perspective. Many officials in the DOE were fully conscious of the need to have a coherent departmental and national agenda. But, at the same time, many were frustrated and at times bewildered by the trials of planning in a political setting. They knew it was necessary to keep trying but often had to hold their noses in the face of the rampant incrementalism that resulted.

Two months prior to the formal launching of the new department, the deputy minister, Robert Shaw, met with his senior officials at Montebello, Quebec, to discuss the DOE agenda and also to set out in a public document what the goals of the department would be.[3] The departmental goals, listed in order of priority, were to be to:

- carry on established resource programs and services;
- clean up and control pollution;
- assess and control the environmental impact of major development;
- initiate long term environmental programs;
- promote and support international environmental initiatives; and
- develop an environmental information and education program.[4]

At these meetings, the essential divisions of opinion in the department became apparent. As chapter 1 has shown, those who wanted to fashion a department devoted to more integrated resource management lost the argument to those who saw the department as a vehicle whose priority was the continued operation of established programs, with some focus on clean-up and control programs and the development of an environmental-assessment process for projects. Environment was thus to become an add-on to the existing mandates and was not to supplant them.

By the fall of 1971, another process, called the 'Main Chances' exercise, was begun. This was done, in part, to integrate the DOE priorities to its program and vote structure for reporting to Parliament. It was intended to develop a ten-year agenda and was assembled after discussion with the various units under two program categories, an environmental program and a renewable-resources program. It included some quite specific undertakings such as reducing motor-vehicle pollution by 30 per cent and a 50-million-pound annual increase in the yield of indigenous fish stocks. But the senior management of the department would not wholly adopt the resulting list, in part because it had no attached budgetary costs. It was, in short, a wish list, but it did show a department anxious to advocate numerous measures, many of which eventually percolated to the top of the agenda in later years.

In the mid-1970s, the department engaged in what was envisioned as an annually updated ten-year planning exercise. Some of the desire for such a long-term view was spawned by the earlier 1972 Stockholm U.N. Conference on Human Settlements. Canada had been a leader in advocating this conference in the late 1960s. The conference's Declaration on the Human Environment enunciated twenty-six principles and a long-term action plan. By June 1975, Canada had prepared, in consultation with the provinces, its own ten-year action program.

The DOE ten-year action plan for 1975 to 1985 was ready by February 1974.[5] It reiterated the Montebello goals and the Stockholm principles and then set out in detail a set of actions that over the ten-year period would cost over $5.6 billion. The plan also set out seven roles which Environment Canada would play. It would be:

- a national voice of knowledge and information;
- a draftsman of environmental requirements;
- a guide to international and intergovernmental negotiators on environmental matters;
- a manager of the fisheries and migratory birds;
- a cooperater with the provinces in managing the air, water, lands, forests and wildlife;
- an implementer and (occasionally) an agent of enforcement; and

- an overseer of environmental activity in federally controlled or financed activities.[6]

   In many ways, this designation of roles (as opposed to earlier goals) was the most telling part of the document. The words used, cited above, came closer to conveying what kinds of powers the DOE had and how it ranked them. It would be a voice, a draftsman, and a guide. It would only occasionally directly enforce. This was already the voice of a more restrained entity with extremely limited means to attain environmental progress.

   Another feature of this departmental effort at ten-year plans was that it began just as the ethos of Trudeau-era rational planning was dying a rapid death in the rest of the Ottawa system.[7] But the DOE justified its effort, in part because of the Stockholm example, but also because, it was argued, environmental problems simply required a longer-term perspective. They were undoubtedly right in having this temporal view but wrong in not having thought through how it could be matched with a political system that was marching to a different drummer. None the less, despite the frustrations and disappointments, there was undoubtedly some value in having a planning process. Without it decisions and priorities would undoubtedly have been even more incremental than they were.

   In 1978, senior officials pressed for another planning exercise, this time centred around the preparation and publication of a proposed White Paper. An environmental White Paper was never published, but some of the drafts of the proposed paper revealed the new restrained ethos that was emerging.[8] The early optimistic notions of a renewable, resource department had been pruned to a mandate which was to be more sensitive to changed federal-provincial relations, the needs of economic growth, and the push within the business community for deregulation. The White Paper exercise again consumed considerable departmental energy but failed to gain approval at the ministerial level. The failure to actually publish a White Paper was attributable, in part, to political cold feet as to just what to do about the environment and, in part, to the fact that an election year was pending, during which environmental issues were unlikely to be central.

   By the end of the 1970s, the DOE had further reined in its planning ambitions. In December 1979, the planning papers were titled 'major priority issues.'[9] The items were shorter but are interesting as a reflection of how environmental problems were being ranked and sorted out. Listed in order as 'very high priority' were acid rain, toxic chemicals, energy development (in light of the Clark government's commitment to the goal of energy self-sufficiency by 1990), forestry, and Northern development. Under the 'high priority' category were items such as weather/climate and waste management. And under the third-ranked items, those which were merely 'priorities,' were problems such as restoration of the Parks Canada capital budget, migratory birds, flood-damage reduction, and public participation.

Four years later, in the fall of 1983, the last of the Trudeau Liberal environmental strategic plans, required under the then-functioning central budgetary-envelope planning system, revealed only slightly altered priorities, albeit a much shorter list.[10] There was only one class of priorities and in it were included: toxic substances, acid rain, the forest sector, water management, climate change, protection of natural heritage, and the North. But what had changed was the tone and style of analysis of the documents. Unlike the earlier documents, which seemed to contain almost no reading of the political and economic constraints and little sense of its own political constituency, this document showed a distinct political learning curve, albeit belatedly expressed.

There was an explicit recognition of the importance of public attitudes and of public support, of the realities of the federal deficit, and of the need for economic growth. For the first time there was an explicit understanding that the department had to 'increase its dialogue with all segments of the public.'[11] One of the first mentions of the concept of sustainable development occurs, and also a greater assertion of the need to move from remedial to preventative measures. The latter included a statement that the DOE would 'intervene more often with regulatory bodies such as the National Energy Board.' [12]

As the Mulroney government took office and after the department weathered the storm of the Blais-Grenier period, the influences on planning shifted from those that were department-centred to those imposed from above. The Nielsen Task Force study team on the DOE was launched because of the Blais-Grenier affair. The study team that looked at the DOE consisted of twelve persons, six of whom were from the non-government sector, including knowledgable persons such as George Layt, a vice president at Stelco, and environmentalist David Brooks.[13]

Although critical of the department, the team also turned its guns on the federal government as a whole. Its first recommendation was that there had to be 'a stronger commitment of the federal government to environmental issues and a more effective role at the federal level for the

Department of the Environment in addressing these issues.' [14] Those directing the Nielsen study teams were not pleased because, unlike other task-force studies, it did not unambiguously focus on ferreting out inefficiencies and expenditures to cut. There were some to be sure, but in effect the report was saying as well that the centre had abandoned the department. As a result, a second study was launched, and its recommendations, combined with those of the first study, contributed to the three-pronged strategy already referred to above in our discussion of the Tom MacMillan era.

By early 1987, as the department's briefing book for the minister showed, the department's strategic thinking was both more realistic and at the same time more sophisticated about the evolving nature of the DOE agenda.[15] The priority list focused on four items: better management of toxic chemicals; acid rain, federal water policy; and the environment and the economy consultation strategy. The department now explicitly acknowledged that there were three kinds of environmental events, and hence three kinds of priorities to be confronted: those which can be predicted and understood; those which cannot be predicted but can be understood and the unexpected ones which can neither be predicted nor understood.' [16]

The department now saw its all-encompassing goal as being one that would 'ensure that the environment is seen by all Canadians as the fundamental support system for all human activities, including economic activities.'[17] The priorities were linked to, but not unthinkingly driven by, public-opinion data, and the document ended with an elaborate chart-based discussion of what the DOE's clientele and interest-group structure was and how the department could reach each component part.

The Department and the Green Plan

As discussed above, the DOE's priority and planning process was galvanized in the post-1988 election period, when the new Mulroney–Bouchard alliance supplied the necessary political muscle. After Len Good and his economic team arrived, senior DOE officials agreed that what was needed was a framework within which all environmental issues, from parks to garbage, from water quality to global climate change, could be situated. It would be a framework within which all existing environmental-policy initiatives could be fit and new ones introduced. It would be, ideally, the sort of umbrella under which all Canadians would pursue their personal or corporate environmental agendas. This was easier said than done.

A small team of DOE officials spent July and August 1990 preparing such a 'framework' document, which the minister was to present to his Cabinet colleagues on the Priorities and Planning Committee on 13 September 1989. The Cabinet approved Bouchard's proposal to proceed with an broad environmental program. But, in a highly irregular step, the day before the meeting, Bouchard announced to the press that a five-year Green Plan would be ready by the time of the February 1990 Budget. This implied that the Green Plan would be an expensive budgetary item, possibly containing tax measures, and would therefore have to be treated like a budget document with all the corresponding conventions of budget secrecy, including limited internal consultation within the government. It also created a perilously brief deadline of about four and a half months for the creation of a document that was system-wide in scope and would therefore require extensive interdepartmental collaboration.

Accordingly, the DOE green-planners who had recently joined the department had to begin crafting the document without much of a chance to even meet with the departmental veterans, let alone consult with the rest of the Ottawa system. When the first request for program content went out, the response was underwhelming. Not surprisingly, the DOE had neither the orientation nor the self-confidence to be leading a major government-wide initiative. In short, neither DOE veterans nor the other government departments believed in late September that the Green Plan would materialize.

Once they became convinced, however, DOE services – the house divided portrayed in chapter 1 – and other departments came forward with scores of program proposals. This creation of a wish list generated an initial cost of around $10 billion. This figure was leaked to the press, which simultaneously raised expectations in the public and upset those in Finance and other key departments who did not want to see an approach taken which would make money the centre-piece of the program.

At this point, however, the analytical framework of the Green Plan was still the need to change societal decision making in order to ensure that environmental damage did not happen in the first place. The focus on decision making meant that responsibility was shared with other societal actors such as business, provincial governments, and individual citizens. It allowed for a forward-looking 'anticipate and prevent' approach as opposed to the clean-up 'react and cure' approach that had been dominant since 1970. Moreover, expenditures were just one of several governing instruments which could be used to change social-decision making. Nevertheless, important central agencies like Finance, which had spent the previous six years on deficit reduction and severe expenditure control, focused on the financial implications of the great green juggernaut.

To control the problem of leaks, DOE officials restricted other departments' access to the document, and this served to annoy and upset officials from other departments. Charges of secrecy from within the system were amplified by charges from both business and environmental groups that they were not being adequately consulted. The emerging tradition of stakeholder consultation that had been built up around the CEPA process and other recent initiatives was being jettisoned. The green-planners who were new to the department were not part of that tradition, and both industry and environmental groups complained that the briefings they were given on the Green Plan 'were not consultation.'

Conclusions

The two halves of the green agenda–setting process, one driven by the episodic preferences of sixteen ministers in twenty-two years and the other by successive efforts at departmental planning, were not entirely separate realms of activity. They fed into each other to produce a particular Canadian brand of environmental priorities. But priorities and ways of viewing things were also influenced by what was going on in the rest of the Western world, be it the Stockholm Conference, the work of the Brundland Commission, or the need to prepare for the Rio Earth Summit.

The sheer high rate of ministerial turnover in the DOE was hardly conducive to consistent or convincing priority setting. The changing and more interdependent character of environmental problems and their considerable unpredictability also meant that DOE ministers and senior officials deserved some sympathy in their efforts to guide the greening of federal policy. Obviously some priority items such as acid rain remained high on the list, while others moved up and down like pistons in an engine. And still others never made it at all.

But, a review of the whole twenty-year period shows that, by the early 1990s, with the forging of the 1990 Green Plan, the DOE seemed to have a somewhat surer and more accurate sense of itself. There seemed to be less of a gap between the first system of agenda setting, the one seen through the interaction of ministers and deputies in the first part of this chapter, and the second, seen through the paper flow of planning and priority lists. There is invariably some gap between these two worlds, but even given this fact, the latter period saw the gap between fact and fancy usefully close. The thinking about priorities obviously also reflected changes in the larger world, from the prosperity of the early 1970s to the 1982 recession, and from the last Liberal days to the Conservative fiscal restraints of the Mulroney era.

But even this account of green-agenda setting is far from complete. Ministers were not just interacting with their own bureaucracy and learning on the run. As the next chapter shows, they were being chased and pursued by other players in the Ottawa political structure, who, for the most part, were opposed to whatever agenda the DOE was pursuing, including the Green Plan.

3

The DOE and the Ottawa System

In the Ottawa system – a mélange of relationships between the centres of executive power and the departments, and among departments and ministers – the Department of the Environment (DOE) had to wend its way through essentially unsympathetic terrain. Indeed, as we saw in chapter 1, Prime Minister Trudeau's explicit vision of the DOE in 1971 was that environmental issues would have to be fought out *among* ministers as well as *within* a single department. The inherent nature of the DOE mandate was that it would have to involve itself continuously in other people's business. But, if it was to have this horizontal reach across the Ottawa system and across its environmental-policy sector, it would have to have, or be given, organizational resources and political capacity in many forms, through friends in high places, through budgets, through scientific and technical capacity, and through support from other departments and their ministers.

Thus understanding the Ottawa system is important for several reasons. First, it is a political and administrative reality in its own right that no political entity can avoid. Second, it is an arena of policy life which, however much it might be portrayed as merely bureaucratic infighting, is in fact primarily a continuous struggle over quite specific resources and political values, not the least of which is the struggle between environmental and developmental values. And third, the struggles and political minuets within the Ottawa system are usually at least in part a reflection of the larger interests of Canadian political life, playing themselves out in the labyrinth of the state and its agencies.

Our account of the DOE's relationships with the Ottawa system is organized in five sections. First, we look at the DOE's dealings with the centres of executive power and its search for political support from entities such as the Prime Minister's Office (PMO) and the Privy Council Office (PCO). Second, the chapter chronicles the DOE's key budgetary battles and its need for financial resources from the Treasury Board and elsewhere. Third, we analyse a selection of the DOE's relationships with other departments. In this chapter, we look at three departments, Indian Affairs and Northern Development (DIAND); Energy, Mines and Resources (EMR); and Industry, Science and Technology Canada (ISTC) and its predecessor industry departments. These are obviously not the only departments examined in the book as a whole. Other chapters examine relations with departments such as Finance and External Affairs.

The fourth aspect of the DOE's functioning in the Ottawa system centres on the department's scientific, technical, and policy-advisory capacities. In the main, we are interested here in their role as a currency or base for the DOE's influence within the Ottawa network. But we also relate this larger analytical and knowledge-based role to the internal operations of the department.

Last, but hardly least, we look at the particular central agency and interdepartmental dynamics involved in the forging of the 1990 Green Plan, without doubt the single-most important policy event in which the political centre both supported and simultaneously disciplined the DOE.

Searching for Central Support

We have already seen from previous chapters that the interest of the political centre in environmental matters essentially involved an early burst of support and resources in the 1970-to-1975 period, a decade-long lack of interest from about 1975 to 1985, and then renewed interest in the latter part of the 1980s and early 1990s. But within these peaks and valleys there were numerous episodes which illustrated that political support and damage could often be quixotic, intended, and unintended as the inevitable jockeying of federal priorities occurred.

No department's fate can be understood without a basic reference to where it stood at various times in the government's priority list. Some of this was revealed in chapter 2, but the essence of the larger priority story is to be found in federal throne speeches. Along with budget speeches, throne speeches are the most telling expression of whether a policy issue is even in the consciousness of a political party in power. In the case of environmental issues, the pattern is quite clear.[1] Throne speeches mention the environment in the early 1970s and the late 1980s. But, in the period between, the political well is dry. Only in the 1980 Throne Speech is there a mention, when it is indicated that mandatory fuel-efficiency standards will be introduced.

This, of course, does not mean that no other occasions arose when the prime minister or other senior ministers expressed concern about environmental issues. But these were essentially mini-episodes in the total scheme of federal priorities. One such instance of considerable symbolic importance in the mid-1970s was the appointment of the Berger Commission of Inquiry into the Alaska Pipeline.[2]

While the appointment and later work of the Berger Inquiry was arguably the best-known environmental event of the latter 1970s, the Department of the Environment had little to do with it. The appointment of Justice Thomas Berger in 1974 was an intervention from the Prime Minister's Office and reflected a concern as to how to respond politically to growing environmental controversy over the building of the pipeline through areas of the North that were not only ecologically sensitive but also involved important Native Peoples' land claims and views of local economic development. The decision on the Berger Inquiry also involved the Department of Indian Affairs and Northern Development, whose minister was then Jean Chrétien.

Berger was seen by Chrétien as someone who, because of his earlier decisions on Native cases, would have credibility. But at the same time, he was thought not to be a risky appointment because he was known to be ambitious for a future Supreme Court position. Both Trudeau and Chrétien were shocked by Berger's ultimate decision to recommend a ten-year moratorium on the construction of the pipeline. But they readily took credit for Berger's innovative hearing processes, which took him into virtually every Native village in the pipeline's proposed path. The Berger process became the quintessential pitting of environmental and Native interests against the oil-and-gas industry. While the DOE was, remarkably, a bystander in the entire Berger drama, it did profit in its wake. Afterwards, the DOE was able to argue that environmental-assessment processes had to be more regular and credible. The government could afford no more Berger-style 'road shows' or surprises. It needed a more open but also a more controllable environmental-assessment regime.[3]

Because decisions regarding the reorganization of departments were in the purview of the prime minister and the Privy Council Office, the DOE's major reorganizations by definition involved initiatives from the centre. In almost all cases, however, the DOE had little influence over the direction of the decision. As shown in chapter 2, it was Prime Minister Trudeau who insisted in the mid-1970s that Fisheries remain a part of the DOE. But it was also Trudeau who gave in to the relentless pressure of Romeo LeBlanc and in 1978 allowed the Department of Fisheries and Oceans to be formed. In 1979, the decision to add Parks Canada to the DOE was not one that the DOE actively sought. Instead the initiative came from Prime Minister Joe Clark and his key advisers acting on a Conservative party caucus undertaking that greater attention might be paid to Parks if it were moved from its current 'Northern' domain in the DIAND to a 'Southern' department. Clark, moreover, had himself been a former Environment critic when in opposition and was a protagonist on park issues because Jasper National Park was located in his Alberta bailiwick.

Three other examples illustrate the varied motivations for the interest of the political centre in environmental matters. First, for much of the 1980s, the political centre, in this case ranging from the Prime Minister's Office to the Department of External Affairs, was politically interested in the acid-rain issue. This issue, however, was often not seen just as an environmental issue. It was also seen in the context of bilateral Canada–United States foreign-policy relations. In both the Trudeau and Mulroney periods, the acid-rain issue was, in a sense, the flagship of Canadian political and symbolic toughness with the United States. This flagship status was especially important in Mulroney foreign policy because, in other respects, the government's critics were charging that the prime minister was being far too cosy with the Americans on issues such as free trade, energy, and foreign investment.[4] This is not to argue that there was no genuine concern at the centre about acid rain as an environmental problem. Rather, it is simply to stress that the motivations for initiatives were multifaceted and crossed over more than one agenda.

The Blais-Grenier episode also involved interventions from the centre of both a planned and an unplanned nature. First, her appointment at all by Prime Minister Mulroney can be seen initially as an expression of the DOE's low-priority status in 1984. Simply on the grounds of Blais-Grenier's total political inexperience, one is entitled to conclude that the prime minister was not putting environmental matters high on his priority list. Later, after the Blais-Grenier débâcle took hold, the political centre had little choice but to intervene. As we saw in chapter 2, the Blais-Grenier affair contributed to the launching of special Nielsen Task Force studies so as to resurrect the department from its malaise. At this time, the PMO was very conscious of the evident rise in opinion polls of the concerns of Canadians about environmental issues.

A final example of supportive activity from the centre was the encouragement given to support the Brundtland Commission. Again both foreign policy and environmental motivations combined here, much as they had in the early 1970s when the Stockholm initiatives were pre-eminent. None the less, the PCO and PMO urged all departments to appear before the Brundtland Commission, and the government fairly quickly committed itself afterwards to Brundtland's concept of 'sustainable development.'[5]

Budgetary Battles

Our acounts in later chapters of the acid rain and South Moresby decisions show further that the nature of central political support was episodic and often unpredictable for the DOE. The same cannot be said, however, for its battles over financial and personnel resources. Budgetary dynamics were best characterized as an intense process of trench warfare, with the DOE continuously on the defensive but, at the same time, occasionally gaining from unrelated budgetary good luck. Its capacity to gain or protect its resources is best seen through an account of its numerous budget-cutting exercises, its relations with fiscal authorities, and the general constraints it faced on resource planning.

In its early years, the DOE enjoyed buoyant budgets, in part because environmental issues were high on the priority list, and in part because Ottawa's coffers were in general quite flush.[6] But from the mid-1970s on until its 1990–1 heyday with the $3-billion federal Green Plan, the DOE was on the defensive. One of the most conclusive forms of evidence of this fact occurred in the summer of 1979, when John Fraser and the newly elected Clark Conservative government requested a review of the DOE and general government-spending practices. In a 28 June 1979 memo to his minister, Deputy Minister Blair Seaborn set out what the department's financial situation had been in the last half of the 1970s.[7]

Seaborn indicated that the department had been 'tightening its belt' regularly and that it was now running most of the same programs, plus several new ones, with 88 per cent of the resources it had had in real terms in 1975. He indicated that the department had experienced during this period four budget cuts: $2.3 million and 73 person-years in 1975–6; $5.0 million and 171 person-years in 1976–7; $6.2 million and 16 person-years in 1978–9; and $33.6 million and 350 person-years in 1979–80.[8] He also stressed that the DOE had: provided for two major diversions of resources to what was now the Department of Fisheries and Oceans; absorbed new programs approved by the Treasury Board but without new resources; and gone through a tough review of its budget in order to obtain resources for high-priority items, such as those needed to enforce the extended 200-mile fishing jurisdiction.

Seaborn conceded that there had undoubtedly been a need for greater efficiency and, in that sense, that some of the cuts were beneficial. But he was concerned that any further cuts would seriously affect service delivery or would indeed lead to the cancellation of programs. Later that summer, in the face of Treasury Board demands for new cuts, Seaborn outlined what kinds of cuts would occur under the requested scenarios of a 5, 10, and 15 per cent cut. He told Fraser that the department could manage a 5 per cent cut but that at the 10-to-15 per cent level, programs would be dropped, which would be of 'major concern to various publics,' and there would have to be 'a significant realignment of federal-provincial responsibilities.'[9] With respect to cuts, he also warned that the department should 'guard against a further erosion of the federal government's research base.'

Further formal budgetary cuts occurred in 1986–7, 1987–8, and 1988–9 as well as in the famous 1985 Blais-Grenier episode.[10] Thus, throughout the period, the department was unable to obtain reasonable control of its resource situation. It did manage to benefit from extraneous financial pools created in the government for other policy reasons. Programs such as the Unsolicited Proposals program and the Special Recovery projects pool of the early 1980s provided some selected short-term relief.

By 1988 the DOE's resource situation revealed a department which, despite facing heightened expectations and demands, was on a very tight financial leash both in constant-dollar terms and in personnel.[11] There is little doubt, however, that the DOE was perceived by fiscal authorities, especially the Treasury Board, as being a financially flabby entity. The Treasury Board obviously has a jaundiced view of most departments in Ottawa since it is its job to say no and to ferret out inefficiencies. As one long-time Treasury Board budget analyst put it, '[The] DOE was for too long filled with too many people who thought they could practise pie-in-the-sky environmentalism, where everything was possible.' But the Treasury Board's capacity to bully the DOE into numerous rounds of cuts was undoubtedly the product of at least two further factors. First, the board knew that environmental issues simply were not a high priority for most of these years and took its cue accordingly. Second, the DOE was not seen to be a good economic department. It was not strong in economic competence either through the presence of economists and economic literacy or through the attraction of upwardly mobile, skilled bureaucractic players who could arm-wrestle their way through the central corridors of fiscal and political power. As for the DOE's musical-chairs brand of ministerial turnover traced earlier, this too was a fatal attraction in the budgetary embrace.

Other Departments

In the Ottawa system, there are always more of them than us. In short, departments, like markets, face their own surrogate form of competition for resources, for political attention, and for room for manoeuvre in carrying out their mandate. This is especially true for those departments such as the DOE whose mandate explicitly involves a horizontal role on policy matters and which accordingly involves a policy sector, as defined in the introduction to this book. Throughout the 1970s and 1980s, the DOE made periodic forays into the interdepartmental wilderness to advance and to test the outer limits of this role. But it also had to be responsive to the initiatives of other departments which it thought might be detrimental to the environment.

A look at the relationship of the DOE to three departments will serve as an initial glimpse of these dynamics: Indian Affairs and Northern development (DIAND) in the mid-1970s period; Energy, Mines and Resources (EMR) in the late 1970s and early 1980s period; and Industry, Science and Technology Canada (ISTC) and its predecessors in the 1970s and 1980s. The DOE's relationships with the DIAND and EMR are especially useful for gauging the flavour of bureaucratic conflicts over classic resource-development issues, and relationships with ISTC capture many of the conflicts surrounding industrial development and competitiveness as economic globalization proceeded apace, especially in the 1980s and early 1990s.

Northern Development Imperatives

The division of the DOE's mandate from the DIAND's in the mid-1970s is a prime example of the constraints faced by DOE officials in the federal government.[12] The DIAND was structured around three program areas – Conservation, Indian and Eskimo Affairs, and Northern Affairs – each of which had a different working relationship with the DOE.

The DOE's relations with the DIAND's Conservation Program were generally good because the DOE had a similar outlook on important conservation issues and was seen by it as a service agency. Conservation, which was essentially the Canada Parks Service prior to its becoming part of the DOE, relied on the DOE for research and advice on renewable-resource and environmental-protection issues. The Canadian Wildlife Service, for example, provided wildlife research and management advice, and conducted other specialized research services under contract with Parks.

Indian and Eskimo Affairs similarly used the DOE as a service, but in different ways. The DOE was used for research and advice on matters of indigenous trapping and hunting, commercial and domestic fishing, and renewable-resource development projects. More importantly, however, the DOE became a major ally in defence of the rights of Native Peoples to subsistence harvests of migratory birds. The U.S. government and hunting groups argued that the Native harvests of ducks and geese in Canada's North were responsible for the population decline of these birds, a position that proved to be untenable in light of the research done by the Wildlife Service.

Relations with the Northern Affairs Program, however, were not nearly as benign. As far as the DIAND was concerned, administration of Canada's Northern territories was the core of its mandate and was only its responsibility, in conjunction with the territorial governments. Northern Affairs was a quasi-provincial government service, and many of the DOE's conflicts with this DIAND Program took on the character of federal-provincial jurisdictional turf wars.

Northern Affairs possessed its own horizontal powers, and was responsible for coordinating the activities of federal government departments in the Yukon and the Northwest Territories. It was also responsible, in conjunction with the Ministry of Transport, for the administration of the Arctic Waters Pollution Prevention Act. These characteristics contributed to continuous friction whenever the DOE tried to act within its mandate on a national basis. Many of the most pressing environmental problems were found in the fragile ecosystems of the Northern territories, and on these issues the DOE usually ran into a stone wall.

The constraints that DOE officials experienced regarding the DIAND were addressed directly by one confidential DOE document that noted: 'There is some degree of coordination with [DIAND] required in connection with virtually every facet of all DOE programs ... Working relationships with the Northern Affairs Program are, in general, not satisfactory.'[13] The DOE thus encountered the DIAND's development ethos across its entire mandate.

On these matters, the DOE was successfully marginalized, in part because of the signals the DIAND was receiving from the political centre. In the mid to late 1970s, a very high priority was being placed on development in the North, and the DIAND had established a very elaborate committee system to coordinate federal activities on a range of issues, many with important environmental implications. DOE management argued that these 'taxed [the] DOE's ability to contribute.'[14] The end result was continuing concern about 'the weight given to DOE input to the various committees.'[15] One senior DOE official described the situation by stating that the 'DIAND's relations with us were a constant struggle because they wanted to run their own show and we could usually do little about it.' As the case-study on environmental assessment (chapter 9) will show, these realities were especially strong when development projects were about to be assessed.

EMR and the Energy Juggernaut

The Department of Energy, Mines and Resources (EMR) evolved in the 1970s and 1980s with the ups and downs of the world energy crises.[16] From the time it was formed in 1966 until 1973, it was largely a technical department, much like the early DOE was. Centred on agencies such as the then-120-year-old Geological Survey of Canada, it was rudely brought into the energy-policy world by the first energy crisis in 1973. We have already seen something of its early relationships with the DOE in chapter 1, where it was noted that, in the debate on the formation of the DOE, there was a significant body of opinion in some parts of EMR regarding the need for a department of 'environment and renewable resources,' based on technical excellence and information. This kind of benign mutual sympathy survived and even supported the early transfer of water-resource units from EMR to the new DOE. The Geological Survey at the time developed within its Terrain Sciences Division an environmental-geology competence intended to complement the DOE and, indeed this action relieved the DOE of the need to hire geologists of its own.

But much of this changed after the first energy crisis, when EMR's policy functions were strengthened and EMR became even more unambiguously a department *for* energy development. As Canada's conventional energy reserves declined in the midst of worries about escalating oil prices and security of supply, attention turned to large resource projects in the North and offshore. The main arena for environmental-resource issues became the fabled Canadian penchant for 'mega-projects.' The previously discussed Berger Commission process symbolized these early confrontations. The DOE had, certainly from the outset, urged the need for environmental assessments, but its procedures for assessments were less developed than they later became and its clout at the height of these energy crises was virtually nil.

Indeed, as noted above, most of the DOE's relationships during the 1975-to-1979 period on energy projects were as much with the Department of Indian Affairs and Northern Development (DIAND) as with EMR. While, by this time, the first energy crisis had subsided, there were still many projects queuing up for federal support. Accordingly, the DOE focused its small-calibre energy–environment policy guns on the DIAND. Although it was still difficult for the DOE to take on a fellow federal department frontally and in public, the DOE did begin to assist some of the environmental groups who were challenging these projects through the improving environmental-assessment processes. The assistance most often came in the form of technical information sent to the environmental groups by sympathetic and concerned DOE technical officials at several levels of the department.

However, in 1979–80, EMR reached almost a mythical juggernaut status in Ottawa, culminating in its lead role in the Trudeau government's highly interventionist National Energy Program (NEP). This pinnacle of EMR's influence coincided with, and was caused by, the second energy crisis. Once again, price rises and security of energy supply became basic political concerns, as well as how the enormous energy rents would be shared between Alberta and the federal governmment. The NEP precipitated an intensely bitter relationship between Alberta and the federal Liberal government, but the policy was also resented greatly by Western Canadian politicians as a whole. This legacy of bitterness was immediately important in itself, but it also returned as a factor in the late 1980s, when the Green Plan process was launched. This delayed effect occurred because many in Alberta, in particular, saw the Green Plan as 'another NEP,' in short, another federal invasion into energy policy.[17]

The NEP was a mixed bag for the DOE. On the one hand, the NEP was the ultimate expression of a pro-development ethos which rolled through Ottawa with little thought as to what other departments or ministers thought. There were certainly some references to environmental concern in the NEP documents, but these were not followed up with concrete funding at the implementation stage. Indeed, the NEP invited, through massive incentives, a wholesale exploration of oil and gas reserves on the Canada Lands in the North and offshore. But at the same time, the NEP, to the extent that it fostered differential and preferential pricing for gas versus oil, began to contain some of the policies that could be considered to be mildly conservation- and renewable-energy-oriented. In that sense, the NEP was pro-environment as well. There is no doubt which was the stronger imperative in EMR, but the NEP did reveal some learning by osmosis. The DOE can take some credit for the EMR learning curve, but it was also attributable to other factors, including the massaging effect of the lobby by environmental groups over the years.

Following its heyday during the NEP, EMR became a very different department under the Mulroney Conservatives. The NEP was jettisoned, and the deregulation of oil and gas began under a pro-market Mulroney government anxious to build bridges with Western Canada.[18] Concerns arose again that environmental issues would be given short shrift at EMR. Interestingly, however, the demise of the NEP and the later sudden fall in energy prices in 1986 provided an unexpected moment for policy reflection at EMR. This occurred in the form of the Energy Options exercise. Designed to cast widely for energy options, the Energy Options committee, headed by investment dealer Tom Kierans, became a forum where, for the first time, environmental groups became a key part of the policy-consultation process on energy matters. The fact that this occurred can in many respects be credited to the DOE in the sense that its own earlier Niagara process meetings (examined later in chapter 5) had begun to pave the way for a modicum of reconciliation between key environmentalists and former ardent energy antagonists.[19]

A final curious point warrants mention about the dynamics of the DOE–EMR play off. This concerns nuclear power. To many environmentalists, nuclear power and related issues of uranium mining were the ultimate environmental cause.[20] Here the DOE appears to have followed an unspoken rule of thumb, which one official summed up as 'Don't even try to get too publicly involved in that quagmire.' In part its historical reluctance to engage itself in matters of nuclear policy was a product of the fact that such matters were in the hands of two independent entities, Atomic Energy of Canada Ltd (AECL) and the Atomic Energy Control Board (AECB), the latter being a health-and-safety and environmental regulator. At the level of technical officials, however, the DOE was involved in providing environmental information to the AECB under the provisions of a joint memorandum of understanding. But beyond this factor, the great DOE public silence on nuclear issues appears to have been based on a continuous political judgment that this was just too much of a hot potato and better left to be juggled by others. Those in the DOE concerned about nuclear issues simply did not see nuclear power as being a black-and-white, pro-versus-anti environmental issue. Given the rest of the issues the DOE faced, this political and technical judgment was probably correct.

ISTC and Industrial Competitiveness

The dynamics of the DOE's relationship with ISTC and its various predecessor industrial and regional departments are very different from the EMR dynamics. Here again, a capsule organizational history is imperative. This history is essentially one that, in the 1970s and 1980s, revolved around how to accommodate the twin pulls of Canada's efforts to devise industrial policies – namely, the reduction of regional disparities and the development of internationally competitive industries.[21]

Until 1982, the regional policy realm was the preserve of the Department of Regional Economic Expansion (DREE), which functioned mainly through grants and federal-provincial regional-development agreements. The industrial-competitiveness mandate was the task of the Department of Industry, Trade and Commerce, whose focal point was the knowledge of its industry-sector branches and its trade commissioners (though the latter was transfered to External Affairs in 1982). Between 1982 and 1987, the twin aspects of policy, regional and industrial, were merged in one department, the Department of Regional Industrial Expansion (DRIE). Also established in 1981–2 was a new overall coordinating Ministry of State for Economic and Regional Development, which reviewed the budgets and policies of some seventeen federal economic departments, including aspects of the DOE's budget.

With growing dissatisfaction over this warring within a single department, and because of the imperatives of a globalizing economy, the Mulroney government again separated the roles. When the Department of Industry, Science and Technology (ISTC) was formed in 1987, it was given a mandate that, to a greater extent than ever before, was to be focused upon international technology-based competitiveness. The government also announced that its new flagship department for the microeconomy was to phase down its use of grants and was to base its role much more on good analysis and knowledge. It was also to become, internally within the government, a reasoned advocate for industrial competitiveness, in short, a more aggressive horizontal agency, much like the DOE was supposed to be in its domain, for environmental matters. ISTC and its predecessor departments had suffered a decline in influence as well, a fact reflected in its own brand of musical-chairs changes in ministers, nine in a decade. But for most of the period being reviewed here, the industry departments did have money. Indeed, on an annual basis, it had probably the largest amount of discretionary funding available in Ottawa.

Given this history, how did the relationships with the DOE play themselves out? Several examples will help flesh out some of these dynamics. In the 1970s, the DREE and Industry, Trade and Commerce (ITC) view was that environmental matters should not be allowed to stand in the way of economic development. ITC was the 'lobby within' in urging a 'go slow approach to regulating sectors such as the pulp-and-paper industry. The DOE reached modest accommodations with ITC when it was able to piggy-back some improvements in environmental production processes as a part of a large grant program to facilitate pulp-and-paper modernization. But this program was basically justified as an economic-competitiveness initiative to enable the huge Canadian pulp-and-paper industry to meet competition from other countries.[22] ITC also lobbied hard from within in the mid-1970s to make sure that the chemical industry was not unduly harmed by the environmental-contaminants legislation. In short, for much of this period, ITC simply defended its most environmentally vulnerable industrial sectors.

As for DREE and the regional domain, its approach was broadly one of attempting to fend off any DOE attempts to devise environmental conditions to the general regional-development agreements that DREE had with the provinces. There were some partial successes, such as in the late 1970s, when federal-provincial forest-management agreements were negotiated between the DOE (led by its Canadian Foresty Service), DREE, and the provinces. But the more pervasive tradition of resistance was reflected under the aegis of DRIE. For example, in correspondence in 1986 between the deputy ministers of DRIE and the DOE, the views of DRIE emerge quite clearly following several meetings with a DOE special task force on environmental-quality policy.

The DRIE deputy minister indicated that 'we are prepared to support the inclusion of references to reflect the importance of environmental considerations in Schedule A of the ERDAs, provided that the provinces agree to do so.'[23] But it chastises the DOE for its failure to retreat from including wording 'aimed at illustrating a clear commitment to balance between economic and environmental considerations.' DRIE regarded this attempt to balance environmental with economic criteria as a 'misapplication of a major federal-provincial economic instrument (the ERDAs) to implement federal environmental objectives.'[24]

The same correspondence reserved its strongest opposition to the DOE when it expressed DRIE's displeasure at the DOE's proposal to assume through Cabinet approval of responsibility for the development of the 'industrial potential of the environmental protection industry.' The industrial sector encompassed the growing number of companies who stood to profit from the goods and services they sold that were a part of the new regulatory demands for environmental protection and for green products. The DOE proposal emanated primarily from Bob Slater, DOE's ADM Policy, in part out of his belief in its inherent importance and in part out of a tactical desire to gain more leverage on other departments. In this instance, the DOE backed off somewhat when DRIE/ISTC agreed to make the environmental-industries sector the subject of one of its 'industrial sector initiatives.'[25]

But the existence of this emerging sector was also telling in the way that it was beginning to change the configuration of coalitions within the newly formed ISTC established by the Mulroney government. This change was partially reflected during an attempted environmental-plan exercise launched in the fall of 1988 when the DOE and ISTC collided in a somewhat different way from the accustomed patterns of the 1970s and 1980s.

Just prior to the 1988 election, Prime Minister Mulroney had committed the government to a major environmental program but one that was as yet undefined. Then, when the election was called, it was left to the Clerk of the Privy Council to ensure that some work was done on this initiative as a part of the 'transition' process that occurs between elections in anticipation of a future new or returning government. The DOE was asked to prepare such a package of proposals. In the course of doing so, a process was devised that was much more open interdepartmentally than was the later Green Plan process of 1989–90. The DOE invited several departments to review proposals.

For our purposes, this autumn 1988 episode is less interesting in its details than in how an industry department like ISTC responded to it. ISTC's basic posture towards the exercise was to resist but also to wait and watch carefully since, unlike most other DOE initiatives, this one was occurring with prime-ministerial and central-agency backing, at least to some extent. But, second, ISTC knew from its own reading of the environmental situation that its own industrial-clientele mix provided a much different political-economic configuration from that of the previous two decades.

ISTC's industrial-clientele sectors now contained not a wholesale opposition camp to environmental issues but rather a threefold division of opinion. One component which ISTC still had to defend was the big polluting sectors, including old favourites such as pulp and paper and chemicals. But a second, smaller group of sectors such as telecommunications and computers included those firms which, reading the same public-opinion signals as was the government in 1988, saw themselves as clean, responsible industries who wanted to distance themselves from the bad polluters. And third, there were the aforementioned environmental-sector industries, a growing and vociferous sector which stood to profit directly from increased demand for their products and services.

While ISTC remained suspicious of the DOE, in part also because the DOE's autumn 1988 proposals were still quite vague, it had to oppose much more carefully both because this was no minor foray by the DOE alone and because the environmental political economy of its own sectors was changing. The 1988 mini–Green Plan process ended abruptly with the appointment of Lucien Bouchard after the 1988 election, but it was indicative of a new era of relations with Ottawa's chief industrial agency anxious to flex its own horizontal powers of policy persuasion.

Scientific and Policy Knowledge as a Base of Influence

While friends in high places and a capacity to muscle and persuade other departments in the Ottawa galaxy are keys to policy-sector power, the DOE from the outset also saw its scientific and technical capacity as a central pillar of its hoped-for influence. This capacity would be vital to feed into appropriate policy advice, regulatory action, the operation of major services, and the conduct of environmental assessments. For most of its history, science (defined as basic research, research and development, and related scientific activities) accounted on average for about half of the DOE's budget. More narrowly defined scientific research involved on average about 15 per cent of DOE person-years. Forty per cent of DOE employees had a professional degree in science or a diploma in a technical subject.[26] These figures exceed those of every other major department in Ottawa.

While the adage that knowledge is power is frequently stated, in reality things are rarely that simple. Yet it is true that, from the outset, many of the DOE's early visionaries saw it as being mainly a scientific and technical agency. We have seen this already in our account in chapter 1 of the debate between those who sought an integrated resource-management model for the DOE and those who sought a narrower regulatory role. But the range of sciences and technical monitoring competencies that would have to be brought to bear for either model would have to be an impressively broad and daunting one, extending from geographers and biologists to experts on engineering and production technologies.

Equally important would be the synthesis and linkage of scientific and technical data to concrete policy problems, where the latter would primarily be in the hands of non-technical personnel. The more that policy itself was involved the more that other kinds of knowledge — political, economic, and managerial — were also needed. In this sense, and despite the fact that they are linked, the story of the DOE's overall knowledge capacity as a base of influence is best told in two stages. The first deals with scientific capacity as such, and the second with how the policy-advisory function evolved.

At the inception of the department, the DOE had in each of its services, except for the Environmental Protection Service (EPS), a solid core of scientists with active networks of contact with the outside scientific community and with active research and data-gathering programs in full swing. Good young scientists were also quite readily attracted to what was viewed as an exciting area of activity. It was this sense of vibrancy and competence that was among the factors prompting early DOE leaders to believe that, on key issues, the DOE had the capacity to 'out-science' its competitors. Case-studies of decisions such as those on Great Lakes water quality in chapter 6 and acid rain in chapter 7 show that the DOE's approach to dealing with the United States was premised on this kind of technical confidence.

Structurally, science was organized and viewed as being a part of, and a support function for, the service, regulatory, and policy activities of the DOE rather than an activity in its own right. There was no ADM for science, nor a separate science service, as existed, for example, in Agriculture Canada. At the DOE, science was a line rather than staff activity. As long as environmental issues could be contained in relatively discrete blocks of service, then the science function could operate fairly smoothly. But from the outset, and certainly by the early 1980s, environmental issues were demanding more integrative science that cut across conventional organizational subdivisions. It was also a science that was far less certain in its outcomes and findings, especially as issues such as global warming emerged.[27]

By the late 1970s, the first serious concerns began to emerge about whether the DOE's early promising scientific start was being maintained. The first impetus for concern was the budget-cutting exercise of 1978. During that year, the axe swung summarily on research programs in forestry, wildlife, and hydrology, and also in support of university-based environmental research. Concerned about the nature and arbitrariness of these cuts, the Canadian Environmental Advisory Council (CEAC) conducted a detailed study of the DOE's scientific capacity and reported its findings to the minister and to senior departmental management in the summer of 1980. The report was very critical, drawing attention, among other things, to 'the remoteness of managerial attention from the emerging environmental issues as identified by scientists' and 'the absence of a department-wide strategy for identifying major issues that required research attention.'[28]

Criticism was by no means confined to external bodies such as the CEAC. Scientists in the DOE, many intensely idealistic about environmental issues and now, in the early mid-career stages of their research lives, were protesting from within. Some did so by exercising the exit option. A senior departmental scientist, who subsequently resigned, circulated a paper entitled 'A Scientific Scream' which argued passionately that the DOE's scientific spirit was being killed. The off-handed dismissal of the paper by senior DOE management merely confirmed a growing view at the technical level that the nurturing of the department's vital scientific capacity was not important.

Many of these criticisms were communicated to the DOE's Science Adviser, Fred Roots. Roots persuaded the senior management of the department to carry out a further review of both the problem of science and that of scientists in the department. While some improvements occurred in the wake of this review, three years later, in 1983, with still further cuts in science budgets, a subsequent paper on the maintenance and utilization of science in the DOE concluded that 'the in-house scientific capacity of the department and its integration with leading science throughout the world is not being maintained.'[29] It also concluded that, even as early as the late 1970s, the department 'found itself unable to apply its scientific resources to urgently needed environmental baseline studies'[30] as urgently requested by DOE regional offices.

It was not until 1984 that the DOE's senior management adopted a departmental policy regarding science and technology.[31] It reiterated the stated view that a science capacity was vital to the DOE mandate but then was virtually obliterated in its effect by the embarrassing and clumsy Blais-Grenier cuts in 1985 to the Wildlife Service's research activity.

The DOE's handling of its science functions has certainly not been exemplary. But if senior management and successive ministers practised indifference, then they were, unfortunately, right on cue with the way science was being increasingly viewed in all political and senior-bureaucratic quarters in Ottawa. While the overall tenor of federal science policy and the attitudes of politicians to science are complex subjects, suffice it say that they were driven by three imperatives, none of which was favourable to the DOE's inherent policy role.[32] The first imperative was that federal policy was driven by a desire to contract out research-and-development activity. A good case can certainly be made for such a policy on a government-wide basis but not necessarily for a department with vital regulatory programs such as the DOE, or for an entity that had increasingly to carry out an especially subtle form of integrative research. The second imperative was that Ottawa's politicians throughout the 1970s and 1980s were mainly interested in the expression of symbolic target setting regarding new levels of industrial R&D (such as the frequently stated target of 2.5 per cent of GNP) without ever delivering the actual resources to get there. The third imperative was that scientific resources were the easiest to cut. Ministers and senior officials often reasoned that science had no viable and vocal constituency, and hence they cut with impunity. Like children with a hose on hot summer's day, they believed that science could be turned on and off at will.

The running down of the DOE science asset base was difficult enough, but it was complicated by a lack of coherence in the DOE's policy-advisory functions, which the science ultimately had to serve. It is virtually impossible to separate out the structural and personnel aspects of policy advising from the inherent political situation that the DOE faced. Thus the DOE was often perceived to have inadequate policy-development structures, in part because most of its ministers were junior and never stayed very long. It was vulnerable because environmental issues simply were not a high priority in the government as a whole, and because the main the DOE services wanted to do their own policy work. Chapter 2 has shown that DOE's chief policy people tried frequently to develop plans that would produce greater coherence in the mandate.

But, at the same time, there is little doubt that the policy function within the DOE was often confused and indeterminate. In the department's evolution from a policy, planning, and research unit in the 1970s to a corporate policy group in the 1980s, the policy function suffered from a significant turnover of leadership and from a struggle for recognition both within the DOE and in the Ottawa network. Given that its greatest critics came from other economic departments in Ottawa, who saw it as an economically illiterate department, one of the more remarkable features of the DOE's policy function was that, until the late 1980s, it refused to strengthen its economic capacity.

This is not to say that the DOE's policy leaders did not have good ideas and a commitment to expressing them. From Al Davidson to Jacques Gerin, to Bob Slater, it had people who ardently and intelligently advanced the environmental cause. But to the rest of the Ottawa system, the ideas seemed too quixotic and intermittent, too frequently unrelated to economic realities, and too uninformed by a coherent view of where the DOE ought to be going.

The Green Plan and the Ottawa System

Many of the characteristics and tensions discussed above emerged in the forging of the 1990 Green Plan. We have already seen the ministerial and departmental genesis of this initiative in chapter 2. But to see how the Ottawa system reacted, we need to look more closely at the contents of the Green Plan.[33] The first point to stress is that the Green Plan is a Government of Canada, as opposed to an Environment Canada, document. It was introduced by the prime minister. Moreover, 60 per cent of the programmatic and fiscal content of the Green Plan falls under the jurisdictions of departments other than the DOE. The plan consists of eight major components, each of which is allocated, in the document, a share of the $3 billion in new funds over the five-year period the plan covers. These additional funds were to be added to the $1.3 billion the federal government already spent annually on the environment, making the government's commitment to environmental spending for the five years roughly $10 billion. However, the ink was no sooner dry on the Green Plan than the Department of Finance, as part of a general expenditure-cutting exercise, unilaterally reduced the five-year total of new money by $600 million in the February 1991 Budget and again in 1992.[34] This reaffirmed the charges by critics that the budgetary commitment of new money involved in the Green Plan was 'soft.'

As we saw in chapter 2, the Green Plan begins with a broad commitment to the concept of sustainable development, calling it no less than an effort at 'planning for life,' the essentials of which are air, water and land. This commitment is then linked successively with renewable resources, special areas and species, and the great fragile Arctic, all of which can be seen as being part of Canada's continental vastness and of Canadians' special environmental image of themselves. Only then are the global imperatives of environmental policy given focus. This is followed by a firm undertaking to change the decision-making process of all political and economic institutions so as to deal with environmental issues with forethought rather than as an afterthought. Emergency preparedness is the last of the topics dealt with, not to give it short shrift, but rather to convey that the old days of reactive clean-ups will soon be replaced by preventative and more systematic approaches.

The first component is titled 'Life's Three Essentials – Clean Air, Water and Land,' and is budgeted at $850 million in new spending. This section includes, *inter alia*, programs on drinking-water safety; a health and environment action plan; funds for river clean-ups; pollution-prevention programs; and programs to reduce toxics, reduce smog emissions, and cut waste in half.

The second component is titled 'Sustaining Our Renewable Resources' and is budgeted at $350 million. Programs here focus on improving forestry, agriculture, and fisheries practices. The third component is on protecting 'Special Spaces and Species' and is allocated $175 million. Programs here are intended to complete the National Parks system, protect Canada's cultural heritage, and protect wild species and wildlife habitat.

The fourth component focuses on 'Canada's Unique Stewardship: The Arctic' and commits $100 million. This includes programs for waste clean-up and management, water-quality improvements, and a research program to investigate the long-range transport of persistent contaminants. The fifth component addresses 'Global Environmental Security' and has a price-tag of $575 million. This includes the various actions, international agreements, regulations, and incentive programs involved in strategies to counter global warming, ozone depletion, and acid rain. It also has a section on strengthening international institutions and advancing international law, as well as furthering bilateral relationships between Canada and individual countries.

The sixth component directs $500 million to changes required to achieve 'Environmentally Responsible Decision-Making.' This is a large component and involves programs to strengthen partnerships between the federal government and the provinces, aboriginal organizations, environmental groups, and industry. Initiatives to improve state-of-the-environment reporting, create innovative national accounting and environmental indicators, and support new science and technology programs are included. There is also a section on legislative, regulatory, and market tools which can lead to improved decision making. The seventh section has a $275-million share and focuses on the federal government's own operations and procedures with the intention of ensuring that they meet or exceed national targets and schedules. Given that the federal government is the largest single operation in Canada – largest landholder, employer, commercial landlord, and purchaser – this is an important section. In addition to employing environmental audits to assess the impact of government operations on the environment and to help develop a federal code of environmental stewardship, this section goes on to commit the government to review the environmental impact of several thousand existing laws, policies, and regulations. This section also commits the government to implementing a new Canadian Environmental Assessment Act and to require an environmental-impact assessment for all proposals coming before the Cabinet for decision. The final section, at $174 million, is focused on improving responses to environmental emergencies, both natural and industrial.

The Green Plan sets out at least some prescribed targets and schedules of action:

- passage within five years of regulations for up to 44 priority toxic substances, and therefore 'virtual elimination of persistent toxic substances';
- a 50 per cent reduction in Canada's generation of waste by the year 2000;
- stabilization of carbon dioxide and other greenhouse-gas emissions at 1990 levels by the year 2000;
- the setting aside of 12 per cent of the country as protected space;
- completion of the national parks system by the year 2000;
- phasing out of CFCs by 1997, and of methyl chloroform by the year 2000;
- a 50 per cent reduction of sulphur-dioxide emissions in Eastern Canada by 1994, capping of acid-rain-related emissions in Eastern Canada beyond 1994, establishment of a national emission cap by the year 2000.

Clearly such a massive initiative was both a threat and an opportunity for other Ottawa players. It was a threat in that it implied a significant incursion into their departmental bailiwicks, both substantively and in terms of how decisions would be made. But it was an opportunity in that a large source of funds was on offer. Departments reacted strategically rather than as environmental converts.

First, the fact that more than half the funds went to other departments was already a sign of the hardball budgetary politics played by other departments prior to the announcement of the plan on 11 December. Second, the Treasury Board and the Department of Finance, strongly supported by other suspicious departments and ministers, ordered that, before Green Plan funds could be spent, the DOE and departments would have to undergo a full review of their existing expenditures. This would give time to root out some of the lower-priority initiatives, and indeed the review process led to a slow pace of announced Green Plan expenditures in 1991.[35]

Not all of the reactions of departments was critical or oppositional in nature. For example, the DIAND was able to garner significant Green Plan money for some badly needed environmental clean-up programs on Native lands. ISTC launched training programs for its own officers and for Canadian businesses to enhance their awareness of environmental decision making. External Affairs and International Trade Canada garnered funds both for the Rio Earth Summit and for eventual support of at least some aspects of the post-Rio international agenda.

But stated frankly, neither the senior-political nor bureaucratic leadership in several departments ever really 'bought in' to the Green Plan concept. While they fought for their departments' 'share' of Green Plan resources and programs once they finally became convinced it would happen, they had not, by the time the Green Plan was announced, or afterwards, really accepted the paradigm shift which implied that things really will be done differently in the future in their departments. Indeed, part of the political achievement was that the government pushed ahead with the Green Plan despite indifference and outright opposition in the Ottawa system.

Conclusions

The Ottawa system, for the most part, did not take kindly to the presence of the DOE in its midst. The political centre periodically took notice in intended and unintended ways, but its overall political support was episodic and rarely strong. This is why the Green Plan stands out as such a startling exception. The budgetary process ground the department down, sometimes deservedly, but mostly because the DOE was politically vulnerable. In its relations with departments such as the DIAND,

EMR, and ISTC, the DOE had some periodic environmental successes, if not directly, then through osmosis and through periodic forays into the interdepartmental wilderness. By the late 1980s, these forays were gaining some momentum, but until the Green Plan of 1990, the DOE could never be referred to as a policy powerhouse. If scientific and technical capacity was to be a further key to the DOE's success in the Ottawa system and in staying ahead of environmental problems, by the end of the 1980s the department, aided by the uninterested political centre, had significantly squandered this vital asset and then tried to rebuild it with Green Plan funding.

In short, the DOE ultimately failed to convince the political centre and other departments that environmental issues were real and threatening their own continuance – that they were not just public opinion and the views of environmentalists. The DOE started the 1990s buoyed by the Green Plan but still facing a phalanx of ministers and policy mandarins who felt that the environment/economy was a trade-off. It had not made a convincing case in the Ottawa system that the environment had moved from an amenity concern to a vital issue.

4

Environmental Federalism and Spatial Realities

The Canadian constitution in its original form (the 1867 British North America Act) did not refer to the environment at all. But, as table 3 shows, when the environmental age dawned, the federal and provincial governments staked their claims under their respective interpretations of the provisions of the division of powers.[1]

The Government of Canada has variously claimed jurisdiction under its authority to make laws for the peace, order, and good government of Canada; trade and commerce; and the criminal law. It has also used its jurisdiction over fisheries, navigation and shipping, and the negotiation and signing of treaties. The provinces have derived their authority in environmental matters from their jurisdiction over the ownership of natural resources, civil law, property and civil rights, and all matters of local concern. While there is certainly considerable constitutional ambiguity on the edges of this vast terrain of powers, there is also little doubt that each level of government knows that the other is in the game for the long haul. They also both know, all bickering to the contrary notwithstanding, that in the long run they badly need each other to help get the environmental job done.

However important the division of environmental powers is, there is also a compelling need to see environmental federalist issues as flowing from the spatial and biophysical realities of air, water, and resources. The character of green federal-provincial relations varies multilaterally and bilaterally not only because of political jurisdictions but also, to a significant extent, because of inherently different problems of air versus water versus toxics and of the particular characteristics of forestry versus oil and gas versus pulp and paper. Causality, of course, does not run in only one direction, from spatial reality to federal-provincial outcome.

TABLE 3
Legislative Powers of Parliament and Provincial Legislatures Relevant to the Management
of Environmental Affairs

Powers of Parliament

- federal lands and Indian reserves (includes administration of any non-provincial territories);
- taxation and spending (taxation powers and leverage through expenditure policies);
- interprovincial and international trade and commerce (environmental control is an indirect result of regulating trade under federal authority);
- census and statistics (information used to influence national standard setting);
- navigation and shipping; seacoast and inland fisheries (indirect vehicles for federal environmental regulation);
- agriculture and immigration (concurrent with provinces);
- criminal law (allows Parliament to legislate the protection of human life and safety; facilitates the development of national standards);
- railways and other works and undertakings of an inter- or extra-provincial nature, and works of multi-provincial or national importance (environmental control is an indirect result of regulating certain industries or transportation modes);
- treaty powers (the federal government is the sole Canadian signatory authority and implements Empire treaties); and
- general powers – laws for the Peace, Order and Good Government of Canada (Parliament can exercise authority under emergency, residual and national dimension doctrines).

Exclusive Powers of the Provincial Legislatures

- public lands; the exploration for, development, conservation and management of non-renewable natural resources and forestry resources in the province; the development, conservation and management of sites and facilities in the province for generation and production of electrical energy;
- property and civil rights;
- local works and undertakings;
- municipal government and facilities;
- all matters of a local or private nature;
- education; and
- agriculture and immigration (concurrent with Parliament).

Source: Constitution Act, 1867, Part VI, 'Distribution of Legislative Powers,' as amended
by subsequent Constitution Acts to and including the Constitution Act, 1982.

The reverse relationship is important as well. But the intent of this chapter is to give greater emphasis than normal to the powerful biophysical determinants of environmental federalism.

In short, federalism is a territorial concept that equates and assigns constitutional and political powers to fixed spatial territories. The environment is inherently spatial and mobile. It is about land, forests, water, and air, which, as interdependent biophysical media, are, for the most part, unforgiving of political boundaries. Hence the conundrum of how to marry space with power. Or, in the language of the double dynamic spoken of in the introduction to this book, it is in discussions of federalism that the desires for control inherent in the first political dynamic of environmental policy initially and most graphically meet the maddening uncertainties of the second ecological dynamic.

We begin an analysis of the DOE's relations with the provinces by tracking the evolution of federal-provincial relations in the realms of, first, water and, then, air. We then discuss more generally the Department of the Environment's experience with managing the federal-provincial relationship as a whole. Next, we trace the DOE's experience with bilateral environmental accords with the provinces. From this kind of account we can then begin to see both conflictual and cooperative outcomes, and how relations and environmental progress varied by province or by regional groupings of provinces. Finally, we look at the 1990 Green Plan and federalism. Several of the vital, but more particular, dynamics of federal-provincial environmental politics and economics will also emerge quite starkly in the case-study chapters on acid rain, the Canadian Environmental Protection Act (CEPA) legislation, and South Moresby.

Water Pollution and Management: Conflict and Cooperation

Water as a medium in which pollution occurs presents two immediate problems, of space and power. Some pollution problems can be 'end of the pipe,' caused by particular plants in fairly localized situations.[2] Other pollution problems can be more broadly area-based, encompassing regions that may be as large as the Great Lakes or the water basin of Western Canada or the Okanagan Valley. Moreover, the nature of the actions involved to end or reduce pollution varies from outright regulation in the end-of-pipe situations to integrated resource management in the larger-area problems.

These characteristics alone helped determine what style of and posture towards federal-provincial relations could occur within the DOE and its various organizational branches. In essence, these relations have typically been more conflictual in the end-of-pipe situations and more cooperative in the larger-spatial situations. In the former type of problem, jurisdiction was more likely to be contested because the water medium was localized, and easily created situations in which firms and plants could utilize their provincial government, especially if it was a regionally powerful industry, as a political ally in any regulatory battles with Ottawa. In the larger-area problems, not only did the water systems usually cross provincial and international boundaries, but the scale of the problem, and the nature of positive as a well as punitive regulatory actions needed, quickly impressed on everyone that cooperation was the key.

But if such configurations of space and power are central in determining jurisdictional dynamics, they are not in themselves a sufficient explanation. Soon, political and organizational imperatives add their own spice to the federal-provincial stew. The fact that federal-provincial relations were both conflictual and cooperative in the area of water pollution can be seen through a brief look at the relations that evolved out of the Environmental Protection Service's (EPS) role in water pollution under the Fisheries Act, and the Inland Water Branch's role.

Although, our main account of the Fisheries Act and the DOE's regulatory role is found in the case-study on the CEPA in chapter 10, three key features of it and of the EPS are essential in understanding the basic federal-provincial posture that the department took. First, in 1971, the EPS was the only truly new organizational part of the DOE. Moroever, public expectations were that this would be the arm of the DOE that would 'get tough' with polluters. Though the EPS was new organizationally, its officers, most of whom came from the fisheries department, and many of these from British Columbia, were the custodians of a statute that had a century-old pedigree. The Fisheries Act had staked out federal jurisdiction over fish, not water per se, and gave the federal government extremely strong, indeed Draconian, powers. The 'new' EPS was in fact an old regulatory brigade that brought with it many of the scars and memories of past jurisdictional problems over the fishery, especially in British Columbia and Atlantic Canada.

The second feature of the Fisheries Act–EPS arrangement was that the Fisheries Act basically applied only to coastal waters. The inland waters and fish were left by agreement to the provinces, in part because they were less important economically, and in part because there were potentially fewer international issues involved. There was also a strong tendency for the EPS to stay away from pollution sources emitted into water from municipal or local sewage-treatment facilities. In the case of Quebec, the administration of key parts of the Fisheries Act was delegated to that province. All of the above features meant that the Fisheries

Act had a very uneven record in the kinds of conflict it induced among the provinces, from very extensive with British Columbia to modest or non-existent with Saskatchewan.

A third element that propelled the federal-provincial dynamics was the leadership of the EPS. Its first assistant deputy minister (ADM), Ken Lucas, was an extremely aggressive man who stoutly defended his organizational territory within the DOE and who also saw the new department as having a tough new regulatory role. Later, as Senior ADM Fisheries and Marine Services, he had little patience for the softer, integrated, cooperative solutions to environmental problems. His EPS became known as the white knight among some industry spokespersons, in part because Ken Lucas wanted the service to be the slayer of pollution dragons. Some of the EPS's early industrial regulatory targets, especially pulp and paper, immediately began to play the federal-provincial jurisdictional card as a way to get Lucas and his crew to back off.[3]

These issues, along with others linked to proposed new habitat provisions of the Fisheries Act led to a major confrontation with the provinces in 1978. This also coincided with the effort by Romeo LeBlanc to separate Fisheries from the DOE, a move actively supported by Lucas, on whom LeBlanc relied greatly for advice. The provincial protest against the new legislation was such that the federal government ordered the DOE to back off its regulatory aggressiveness. This in turn led to an even more pronounced tendency for the DOE to delegate *de facto* compliance and inspection to provincial authorities, though this practice had begun earlier. From 1978 until the changes to the legislation under Canadian Environmental Protection Act in 1987, federal-provincial environmental relations under the Fisheries Act were relatively cooperative, or at least benign.

The contrast between the EPS–Fisheries Act relationship and that which emanated from the inland-waters aspects of the DOE mandate was stark. The prime determinant here was the inherently broader spatial nature of the problem and the need to plan for integrated resource use rather than just to ban unacceptable behaviour. But here too other political and organizational boundaries had their influence.

Consider first the Canada Water Act. It was not a regulatory act but instead a facilitating statute that was intended to induce cooperation through research and other joint initiatives. It was also a new statute that had no pedigree of accumulated grievances and memories. Second, the Environmental Management Service of the DOE (within which was located the Inland Waters Branch) was not organized by industrial sector, as was the EPS, but rather by broad resource elements (water, land, etc.) and was composed largely of scientific and technical people inclined to think more in terms of integrated resource management. Moreover, one of the first major environmental initiatives centred on Great Lakes water quality, a project that required an examination of the Great Lakes in systemic terms.

Some of the ethos for this work on water predated the DOE and was crystallized in the work of the Canada Centre for Inland Waters (CCIW), originally located in EMR. The CCIW had been established as an inter-disciplinary scientific establishment, in part out of the premise that, if Canada was going to successfuly negotiate with the United States over Great Lakes quality, it would have to, as chapter 3 pointed out, 'out-sci-ence' the United States to compensate for the latter's greater industrial political clout. Once located in the DOE, the CCIW, and later the Inland Waters Directorate, used this technical tradition as a base for building federal-provincial bridges.

One of the early areas of relatively good cooperative relations emerged with the early need on the Great Lakes to regulate phosphates in detergent products. While the detergent industry certainly battled with the DOE, there was little dispute with Ontario on these matters. Indeed, early cooperative principles were established with Ontario, in this instance in part because of the willingness of Walter Giles, the senior player for Ontario, to see a natural division of labour between Ottawa and the provinces on this kind of issue. Thus, Ottawa would set goals and standards based on its stronger research expertise, and the prov-inces would implement and deal with local variations. Of course, it helped in this case that the detergent industry was not in itself large. Had it been so, the jurisdictional stakes with Ontario would have been greater.

But detergents were more like a single end-of-pipe problem. Some-what more symptomatic of the larger EMS–water role occurred in areas such as flood control. Key players in Inland Waters such as Jim Bruce had extensive early experience in flood-plain mapping. Early work in the late 1950s conducted in the wake of hurricane Hazel had led to extensive zoning of river valleys and shorelands on Lake Ontario, and indeed paved the way for the establishment of extensive parklands on the shore areas. In the mid-1970s, the Inland Waters people were well on their way to arranging high-water or flood-control agreements with the provinces. This was done in part to ensure that Ottawa was not hit repeatedly with bills for flood damage when areas subject to such flood threats were allowed to be built up with housing and development. Instead, a federal-provincial system of mapping and zoning was secured to ensure proper resource use and to prevent flood damage.

Similar examples of cooperative action occurred in areas such as sensitive wetlands mapping and zoning. Both the flood-control and wetland-mapping approaches developed at the DOE were used in many other parts of the world. Another form of good institutionalized federal-provincial cooperation came in the form of the Prairie Provinces Water Board. It and various provincial water committees were not headline-grabbing initiatives but took advantage of the use of federal DOE bridge-building research activity to achieve significant improvements in water resource use and planning. Thus, in a very strong sense the EMS saw the provinces as its clientele.

Air Pollution: Institutionalized Cooperation

Air is even more mobile than most bodies of water, and thus federal-provincial cooperation in relation to that medium became even more institutionalized, mainly through the Federal-Provincial Advisory Committee on Air Quality (FPACAQ).[4] Federal initiatives on air were partly carried out through the Clean Air Act, which, much like the Canada Water Act, was more a facilitating than a regulatory statute. Until 1982, DOE leadership on air pollution issues came from the Air Pollution Control Directorate in the EPS, after which the directorate was abolished as a separate entity. The Atmospheric Environment Service (AES) was also involved in air issues.

The origins of the Air Pollution Control Directorate (APCD), and hence important parts of its operating culture, were in the Department of National Health and Welfare's Environmental Health Directorate. Thus, the early work on air issues was mainly seen from a human-health perspective. When brought into the fold of the DOE, the health-oriented directorate had been in existence for only two years but had already been active in the formative work on what was to become the Clean Air Act. Health and Welfare, as occupier of a policy field inherently involving concurrent jurisdiction, had been used to operating in a joint manner with the provinces. Thus, at the outset, a federal-provincial committee, initially an *ad hoc* one, was struck in December 1969. Its purposes were to promote better understanding of the nature and extent of air-pollution-control activities in Canada, and to review the desirability of conducting a coordinated and integrated national air-sampling program. In addition, it would explore the issue of national air-quality criteria and standards.

The early committee meetings also established the ground rules for the federal role, to help allay provincial concerns about the expected clean-air legislation. A proposal from Ontario, quickly agreed to by all the players, was that the federal role in air pollution would be: leadership in the development of air-quality criteria and goals; the provision of a coordinating function to ensure some sort of uniformity in the data collection by the provinces; the assembly and distribution of scientific and technical information; the provision of technical assistance to those provinces that could not provide it themselves; and the handling of interprovincial air-pollution problems.

This early work led eventually to the establishment of the FPACAQ and to fairly consistent practices based on these original principles. In 1971, an Air Quality Subcommittee had already begun its work and had recommended desirable and acceptable levels of air quality for sulphur dioxide, suspended particulates, sulphur dioxide in combination with suspended particulates, carbon monoxide, and oxidants (ozone). By 1974, these desirable and acceptable levels had been promulgated under the Clean Air Act, which had come into effect in November 1971. The Air Pollution Control Directorate in these early years had grown from 30 to 100 persons, and there was a genuine sense of achievement and progress, both in general and in the federal-provincial arena.

By 1976, the FPACAQ's role had been scaled down somewhat so as to ensure that it was unambiguously a *technical* committee. A major reason for this change was that the processes of managing and regulating air quality were reaching a new and more difficult stage. In the first round of establishing air-quality objectives, reaching agreement was not difficult on those objectives labelled desirable and acceptable. But the third level, setting tolerable levels, was infinitely more political and economic in its immediate impact. Since in situations where tolerable levels were exceeded, the federal government could step in if the province did not, much slower progress occurred on setting tolerable levels as both provinces and industries dug in their heels.

In the latter part of the 1970s, air-quality levels for nitrogen dioxide (NO_2) were promulgated after the committee's work. But again some further analytical problems occurred. There was controversy over the technical accuracy of the samplings in the United States on which the objectives had been based. As well, some provinces simply did not have the technical capacity to monitor or measure levels. Again, it is essential to stress that, in the latter 1970s, the deteriorating economic situation was influencing the politics of standards setting.[5]

When the Air Pollution Control Directorate was abolished in the 1982 reorganization, the committee's role was reviewed to take stock of the new situation as well as of the changing nature of air-pollution problems. The revised terms of reference for the FPACAQ were based on the essential principles established previously but with the additional proviso that federal-provincial air-quality management had to reflect the fact that air issues were now much more complex and that air was 'a medium by which pollutants are transported to impact other parts of the ecosystem.'[6] These factors made federal-provincial dialogue all the more necessary and all the more difficult.

It is roughly in the early to mid 1980s that it becomes impossible to tell the federal-provincial air-pollution story through the aegis of the FPACAQ alone. Its history and agenda join other policy and organizational trails as air issues became more complex. These include the particular federal-provincial (and international) journey of the acid-rain issue (see chapter 7) and the larger needs to manage and institutionalize the general federal-provincial environmental relationship.

Managing General Environmental Federalism

To say that the general federal-provincial relationship was managed is to evoke a sense of serene control that was in fact not achievable. An accurate account of macro–environmental federalism requires a brief look at several evolving developments and institutions. These include the formation of the Canadian Council of Ministers of the Environment (CCME); the strengthening of provincial environmental capacity; the impact of the cuts in DOE spending and of the withering of its technical base; the impact of public-opinion support for federal leadership; provincial opposition to being considered merely another stakeholder in the DOE's consultative processes; and the consequences of landmark court cases on federal environmental-assessment processes.

The CCME had been spun out of the larger Canadian Council of Resource and Environment Ministers (CCREM) in 1988.[7] Indeed the previously mentioned FPACAQ had felt the need to clarify its relationship to the CCME because more and more of the federal-provincial multilateral action seemed to be in the CCME's bailiwick. Accordingly, it approached CCREM deputy ministers in 1987 to have the CCREM endorse its terms of reference, funding arrangements, and planned activities. This was agreed to, in part out of a recognition that there were genuine problems of overlap. The CCREM, for example, had set up its own air-quality objectives for PCBs but had set up a different concept for the establishment of acceptable levels of air quality for PCBs than that established and understood by the air-quality community.

The CCREM had itself been formed in 1971 as an elaboration of the earlier Canadian Council of Resource Ministers established in 1964. Its mandate went well beyond air-pollution issues and embraced all issues of environmental management, but in the 1980s its Committee on the Long-Range Transport of Airborne Pollutants (LRTAP) became a key working engine for the acid-rain issue and later for the management of issues such as oxides of nitrogen (NOx) and volatile organic compounds (VOCs). Composed of ministers, but supported by a mirror committee of deputy ministers from all eleven governments and a secretariat, the CCREM (now CCME) fostered information exchange, consultation, and debate about environmental issues, leading to cooperative federal-provincial initiatives. The chairmanship of the CCME rotates among the eleven ministers and thus in theory puts all of its members on a somewhat more equal basis in terms of influence and responsibility. In 1989, the CCME set up a permanent and enlarged secretariat in Winnipeg and established two committees that reflected its now even more important role. One committee focuses on strategic planning, and the other on current environmental operations. Among other things, the CCME is a particular catalyst for the establishment of federal-provincial equivalency agreements, as required under the Canadian Environmental Protection Act. The CCME thus reflected in different ways the need to institutionalize the growing scope of environmental federalism.

The general relationship was also influenced by the increasing growth and competence of provincial environmental ministries. Initially, Ottawa had the edge in expertise in many areas, but it did not take long for several provinces to catch up with and, in some sectors, to surpass the DOE. The provinces, partly out of fear of federal incursions, but also because of local political pressure, established their own environmental-assessment processes on projects. Local industries would then frequently buttress their support for provincial assessment processes, in part out of opposition to having potentially to go through two environmental-assessment processes, one federal and one provincial.

But while provincial expertise allowed the provinces to more than hold their own with Ottawa, this capacity was by no means uniform.

While Ontario, Quebec, Alberta, and British Columbia had solid overall capacity, other provinces, especially those in Atlantic Canada, often looked to the DOE for help, in terms of both expertise and money.

In the overall federal-provincial realm, some of the tension always also had a partisan political stripe. Ministers from the party in power would approach each other somewhat differently than they would ministers from opposition parties. There was also always a particular tension between Ottawa and Ontario, regardless of the party in power. In part, this tension was attributable to the power and concentration of the Toronto media and the need felt by federal and Ontario environment ministers to upstage each other, through either claiming credit or assigning blame.

A further factor which altered and weakened the DOE's capacity to influence the provinces in the 1980s arose from the simple fact that the DOE's budgets, as chapter 3 showed, were eroded and its scientific capacity was reduced. The provinces always objected to any federal regulatory power, especially, as we have seen, under the Fisheries Act. But, for the most part, they welcomed DOE money and scientific input. When both these instruments of influence attentuated in the 1980s, Ottawa lost still further clout.

The only counterweight to this situation was that Canadians in most provinces, even when Ottawa failed to act, saw the DOE and the federal government as the more credible environmental leader. The provinces were often perceived as being in too cosy a relationship with dominant local industries such as forestry or oil and gas or smelters and power companies. By the late 1980s, the DOE's public-opinion polls showed that Canadians expected federal leadership on environmental matters.

Eventually the DOE began to act on the assumption of the strength of this extra political leverage. One indirect manifestation of it came in the DOE's evolving stakeholder consultation processes examined in chapter 5. In many of these exercises, the DOE treated the provinces as, and called them, 'stakeholders,' as if they were indistinguishable from any other environmental or industrial lobby group. By the early 1990s, the provinces were strongly objecting to being so designated. They were fellow governments, not mere interest groups.

The general federal-provincial relationship was also profoundly affected in the early 1990s by court cases which forced Ottawa's hand and which compelled a strengthened federal legislative approach in the assessment of major projects. The court rulings on the Rafferty–Alameda dams in Saskatchewan and on the Oldman River project in Alberta are examined in chapter 9. At this point, it is sufficient to say that they brought the provinces and Ottawa to a head-to-head conflict over what the new saw-off between area and power would be.

But the DOE's political confidence had in fact begun to grow in the post-1986 period, and its view of federal-provincial relations began to change accordingly. We saw some of the flavour of this in chapter 3, where we showed how the DOE had become more aggressive in attempting to influence the content of the general regional-economic-development agreements with the provinces.

But it was the process that produced the Canadian Environmental Protection Act (CEPA) that led to the most concrete consolidation of views about the DOE's role in federal-provincial relations. One of the central purposes of the CEPA was plainly and simply to strengthen federal jurisdiction. The federal government had long been stung by the criticism that it had basically left enforcement to the provinces.[8] Some of the desire to assert federal authority also occurred at the prime-ministerial level, since the Mulroney government was increasingly being accused of having weakened the federal government by being too conciliatory towards the provinces in other policy fields. In any event, at the centre of the new CEPA regime was a series of far tougher federal-provincial agreements.

While the larger CEPA story is told in chapter 10, an examination of the nature of the CEPA-style agreements is a fitting close to this brief account of general federal-provincial relations. These agreements, will in some respects, be broader than the bilateral accords discussed below, though they are not holistic in the manner desired by some. But they will certainly be more detailed and tougher in asserting federal jurisdiction. Thus, under the CEPA, deals are being forged on environmental-quality monitoring systems; the disclosure of confidential business information; equivalency agreements on the regulation of toxic substances and, separately, on international air pollution; and on general issues regarding the administration of the CEPA.

There is little doubt that the general federal-provincial domain of environmental policy is extraordinarily multifaceted. The DOE ministers' various briefing books over the years reveal at any given time, and for each province, dozens of ongoing issues, from small to large. Most are resolved as a matter of course. Some become test cases for the practice of intergovernmental environmental warfare or for the studied art of deflecting blame onto someone else.

Environmental Accords: The Need for Practical Bilateralism

While separate accounts of water, air, and then general relationships in the federal-provincial domain are indicative of different kinds of dynamics, there was always in addition, as there has been since the earliest days of the DOE, a more specific bilateral set of dynamics between Ottawa and each province. These can best be seen, first, through a look at the evolution of the DOE's experience with federal-provincial environmental accords.

In 1975, Federal-Provincial Accords for the Protection and Enhancement of Environmental Quality were signed with seven provinces. Newfoundland, British Columbia, and Quebec did not agree to accords. Newfoundland declined, largely on constitutional grounds, as did Quebec. Both regarded environmental matters as a provincial responsibility. British Columbia, on the other hand, did not disagree in principle with such accords. Instead, it said it would not sign an accord until the federal government clarified the respective roles of the DOE and Fisheries in the protection of fish from pollution. Set initially to expire in 1980, they were extended for five of the provinces for a further year, and in 1983 the accord as an instrument was reviewed.

The 1983 review took cognizance of several changes that had occurred since 1975.[9] Among these were the separation of Fisheries from the DOE in 1979; a critical report by the auditor general, who, based on financial grounds, had raised concerns about the arrangements with the provinces for the enforcement of pollution standards; the role of the DOE's new regional directors general, who had reponsibility for the general conduct of relations with the provinces; the changing interrelated nature of environmental problems and priorities; and the then-new DOE Policy on Public Consultation.

Assessments of the effect of the accords in the early 1980s produced a mixed judgment. The accords were intended to establish broad principles for federal-provincial cooperation. They set the responsibilities of the two levels of government, including assigning to the provinces responsibility for enforcing environmental regulations at least as stringent as those required under federal regulations. An internal 1981 DOE study based on discussions with the DOE's regional directors general (RDGs) had identified only limited impacts, 'except where the accords have been an instrument to encourage provincial implementation of federal requirements.'[10] such as occurred with mercury discharges in Ontario. The implementation committees called for by the accords were rarely put in place.

The assessment also brought out the fact that there was a marked similarity in federal-provincial working relationships in both the signatory and the non-signatory provinces. It also showed that the accords were seen within the DOE as being largely concerned with the mandate of the Environmental Protection Service rather than as department-wide instruments. Thus, where they were included, other departmental areas such as 'environmental enhancement,' 'environmental assessment,' and 'environmental design' were not described with sufficient precision to allow concrete action. In general then, the internal DOE view of the accords was that they helped convey some early spirit of cooperation and had some selected areas of success, but overall were not decisive in the federal-provincial relationship. Some, however, were more critical. As early as August 1981, Jacques Gerin, the ADM Policy, had reached his verdict. He stated that the accords 'have demonstrably not been essential to good environmental performance nor have they been conducive in themselves to cooperation or the development of further specific agreements. Their value therefore is mainly symbolic.'[11]

As for the provinces, their view seemed on the whole to be more positive. During the 1983 review, all saw some value in the accords, perhaps because the accords were a small source of funding when money was scarce, but also because they were a vehicle for addressing at least some local environmental issues. Newfoundland and Quebec remained outside the accord framework. As for British Columbia, it had agreed to take part but was having second thoughts until it found out what might emerge jurisdictionally from the federal Pearse Commission's report on water policy.

By the mid-1980s, the DOE's internal preference, partly promoted by its regional directors general, whose function in the DOE reporting relationship was increasingly ambiguous, was for more holistic agreements in order to cover all areas of the DOE mandate. But having agreements coordinated in this all-pervasive way was usually offensive to the line branches, who preferred to deal with their own devices. Indeed, as we have seen in the sections above, areas of the Water and Air components of the DOE felt that they had evolved very good relationships with the provinces.

The variety of issues and the large number of decisions covered throughout the twenty-year period mean that only with great caution, if at all, should one attempt to generalize about particular provinces along the cooperation–conflict continuum, let alone in relation to which province best meets the green test in substantive terms. All provinces have grown in terms of their basic environmental capacity and basic learning capacities. But each does bring its unique political economy to the environmental bargaining table. Quebec asserts its autonomy in an across-the-board fashion but cooperates frequently and where necessary. But it is especially tenacious over James Bay hydro projects. Ontario has been basically cooperative at the bureaucratic level, in part because its economy is broadly based and its environmental offenders are spread around a diversified economy. It also has the most extensive day-to-day contact with the DOE. But at the political level, Ontario sees itself as the main competitor of the DOE for environmental leadership. Alberta is especially protective of any incursions into the oil-and-gas industry, its primary economic base, but otherwise has cooperated in many areas. British Columbia has been overwhelmingly watchful of the Fisheries Act and of various DOE incursions into forestry and water policy. Saskatchewan has sparred over uranium mining and agricultural pesticides, and exploded over the Rafferty-Alameda dams decision. And so the list goes on.

None of these basic configurations should be surprising. These patterns reflect where the natural resources of Canada are concentrated. Resources *are* the environment. Spatial realities and political-economic power collide, sometimes sharply, and on other occasions like the slow but inexorable movement of the geological plates beneath the earth's surface.

The Provinces and the Green Plan

As the 1990 Green Plan came on the agenda, it triggered a classically Canadian federal-provincial battle over both substance and symbolism, green and otherwise. The Green Plan document was remarkably silent about detailed jurisdictional issues, in part because the federal government raised the ire of the provinces because of the way it went about forging the plan in the first place. As mentioned earlier, the provinces were opposed to being viewed as just another 'stakeholder.' But provincial reaction to the Green Plan was also in fact larger than the plan per se. Bundled in with the Green Plan were provincial concerns in 1990–1 about the *Rafferty-Alameda* and *Oldman River* court cases and the new legislated aspects of federal environmental-assessment processes; the practical effects of the CEPA; the cuts in general federal-provincial joint funding, seen by the provinces as off-loading the federal deficit onto them; the critical problems in the east-coast fishing stock; and the imposition of the new federal Goods and Services Tax (GST), followed by burgeoning concerns about a potential green energy tax to combat global warming.

During the fall and spring of 1989–90, most provincial governments and many of their allies in industry wished the whole idea of a Green Plan would just go away. But when the Green Plan was known to have serious political backing in Ottawa, provincial behaviour had to take a dual tack, much like federal departments had. First, the provinces sought to defend jurisdiction and protect themselves against Ottawa's potential or actual unilateralism, especially regarding natural resources. Second, the provinces sought to get federal money wherever they could, playing a new green version of a very old political game.

Thus some provincial critics were pleased that the Mulroney government had committed such a substantial sum of new money to the environment during a period of severe fiscal restraint. Moreover, the Green Plan stressed shared responsibility for the resolution of environmental problems, and identified several areas where greater federal-provincial cooperation would be required. Potential areas of conflict were simply swept away with soothing rhetoric. After its release many provincial politicians were pleased that the Green Plan did not unilaterally introduce any initiatives which would appear to involve Ottawa in provincial areas of jurisdiction. This was a relief to the provinces, which felt that the Canadian Environmental Assessment Act, which was then before Parliament, did encroach on provincial jurisdiction. However, other provincial politicians, who acknowledged that the Green Plan would require a more intensive level of environmental negotiations, complained that the provinces had not been adequately involved in crafting the plan.

Provincial objections to a potential Green Plan in 1989–90 were closely aligned with industry, especially in Alberta's energy sector. Partly because of Lucien Bouchard's unilateral commitment earlier in 1990 to keeping Canada's carbon-dioxide emissions in the year 2000 to 1990 levels, both Alberta and the oil-and-gas industry mounted a fierce lobby against the Green Plan.[12] Alberta interests had feared an interventionist Green Plan along the lines of the 1980 National Energy Program (NEP). Consequently, they lobbied hard, employing anti–NEP-style rhetoric, which played well in Western Canada as a whole. Alberta and energy interests also pressed for an approach which would emphasize market-based instruments and prior consultation about the costs and implications of the effects of new regulations on international competitiveness. Alberta, along with other provinces, was pleased that the Green Plan avoided new carbon taxes and committed the federal government to public and industry consultations before developing further regulations in areas such as controlling greenhouse-gas emissions.

Conclusions

The Canadian constitution ensures that environmental policy is an area of divided jurisdiction between the federal and provincial governments. The DOE's roles and capacities have been affected from the outset by these realities of Canadian federalism. But environmental federalism proceeds as much from a spatial, biophysical imperative as it does from constitutional powers per se. The DOE's approach to dealing with the provinces has accordingly not followed any single path but rather has evolved in a variety of bilateral and multilateral ways. Water pollution produced both 'end of the pipe' confrontation and sustained cooperation in some larger areas of water management, especially when federal financial coffers were full. Air pollution was largely cooperative in the early stages but became more difficult as the nature of problems themselves became more ecologically integrated. Efforts to embrace the entire relationship with any given province evolved from the rather benign use of environmental accords to the more compelling contents of the CEPA equivalency and related agreements, including those that might be facilitated by the Green Plan's funding.

But inevitably, environmental federalism is also driven by the essential features of a diverse regional geography and political economy. Each province becomes a defensive and protective force when the fundamentals of its regional political economy are environmentally challenged by Ottawa too swiftly or too frontally, or with too much implied support from Canadian voters.

5

The Elusive Constituency:
ENGOs, Business, and the Public

Over its more-than-twenty-year history, the Department of the Environment (DOE) has had difficult relations with what proved to be an elusive outside constituency. This constituency consists primarily of the environmental non-governmental organizations (ENGOs), business interest groups, and public opinion in general. The three constituent elements constantly fed off one another. In turn the three elements together poked and probed the department, each in its own way trying to get the department to do its bidding.

Any analysis of the DOE's efforts to manage its constituency, and of that constituency's counter-effort to influence the department, requires a careful and considered approach. For both of the two decades covered, we need to map out the character of the interest being examined to get a sense of the problems faced and of how these problems changed. Second, we have to look at the issues that the DOE faced in dealing with each interest. For the ENGOs and the DOE, the story essentially concerns the problems of how to build a supportive constituency among nominally natural allies who never quite trusted one another. And for business and the DOE, the trail of relationships is one in which industry, with the aid of the provinces, moved from a position of favoured occupant at a small inner table in the regulatory process of the 1970s to one of many stakeholders sitting at ever-larger environmental round tables in the 1980s and early 1990s. And for business and the ENGOs, much of the story consists of parallel monologues, until the early 1990s, when an actual bilateral dialogue emerged, increasingly without government as the intermediary.

The chapter also examines the role of public opinion and of the DOE's efforts to influence it or, occasionally, to ride waves when favourable.

Finally, we look at the Green Plan decision process where all three elements – business, the ENGOs, and the public – played a role in the DOE's and the federal government's most ambitious environmental initiative.

The ENGOs and the DOE: Reluctant Allies

Every federal department has to have or build an outside-interest-group constituency to support its basic goals and to keep up the pressure when agendas are shifting and unfavourable.[1] But the key behaviourial feature of the ENGOs writ large is that, as a political lobby, they are the quintessential example of an interest group that is afflicted by what economists call the 'free rider' problem.[2] This problem involves a situation in which ordinary Canadians are able to benefit from the work of the ENGOs – in short, from the public goods that they produce – without ever having to join or pay fees to the groups. As a consequence ENGOs have problems obtaining and retaining sufficient members to fund the nature and scale of their lobbying operations.[3]

The character of environmental problems in the 1970s and 1980s also had a major impact on the lobbying activities of the ENGOs. Local and regional environmental problems were dominant in much of this period, which encouraged the formation of single- or local-issue environment groups. An ENGO of this type would rise and decline in accordance with the fluidity and fluctuating public profile of the problem it was identified with. Moreover, local environmental activists were compelled to spend most of their time and money on the core issue or problem at hand, leaving little room for establishing lobbying networks with other ENGOs.[4] One prominent Canadian environmentalist noted, 'In the 1970s we would be working away on our local or regional problem and did not even know if there were other groups doing the same in other parts of the country.' The ENGOs were also fragmented by the sheer geographical size and diversity of the country, and this discouraged organizational links between them.[5]

The potential scale of ENGO operations as lobbyist and public educator can further be seen by briefly attempting to visualize the underlying production cycle and physical flows of environmental hazards and problems.[6] Any particular hazard can traverse the full production cycle, from extraction from the ground to fabrication, from hinterland workplace to urban workplace, and from transportation within Canada to shipment to places abroad as products evolve into higher value–added forms. At several points along the production chain, governments already seek to intervene in the name of health, safety, and environmental values. But all of this is for just one hazard or product, and there are literally thousands of them, spread over regional, national, and international jurisdictions. Moreover, they are linked to hundreds of bodies of water, urban and rural land configurations, and air sheds. Little imagination is required to realize that even the most zealous and determined ENGO cannot possibly be in the tens of thousands of sites where environmental behaviour is intended to be changed.

But the flip side of this inherent political weakness, especially in contrast to the relatively greater cohesiveness of business interests, is that the ENGOs are driven by a strong value commitment, and their core adherents are deeply dedicated to the long-term mission. Moreover, precisely because of their free-rider weaknesses, the ENGOs have sought to use the mass media and their links with the Canadian public as their basic counterweight to the influence of business in the environmental-policy process. While the media supply some political leverage, reliance on them certainly does not overcome all the problems associated with the inherent fragmentation of the ENGOs.

ENGO fragmentation must also, in large part, be explained by divergent environmental philosophies. As Glen Toner's analysis shows, four main philosphical strands can be found among these groups in Canada: conservationism, environmentalism, preservationism, and ecologism.[7] Conservationism has had a very long history within Canada and several influential ENGOs.[8] One key aspect of this philosophy that aids in understanding its development is that it was born out of disputes within the political and economic establishment of the nineteenth century. Conservationists, including many prominent community leaders, developed conservative positions in defence of commerce and leisure activities undertaken within wilderness areas, such as sport fishing, hunting, trapping, bird-watching, and touring. The formation of the Royal Commission on Fish and Game, set up in the 1880s and reporting in 1892, and the Federal Commission of Conservation, established in 1909 by Wilfrid Laurier and Clifford Sifton, reflected a growing popular concern.[9]

Since the late 1970s, conservationism has shown up in well-known works by the Club of Rome, specifically the study, *Limits to Growth*, and the Science Council of Canada's research, published in *Canada as a Conserver Society*.[10] It has also dominated within several prominent Canadian ENGOs, such as the Canadian Wildlife Federation, Ducks Unlimited Canada, the Nature Conservancy of Canada, and the Canadian Nature Federation. Each of these groups embraces a mix of philosophical positions, but each has a predominant orientation towards conserving scarce resources through sustainable-use practices.

The second philosophical strand, environmentalism, dates back to the 1960s and early 1970s.[11] Environmentalists focus on removing and mitigating the ecologically damaging practices of modern industrial society. As Toner notes, 'The "first wave" of environmentalism gave rise to a series of pointed and vehement confrontations, as newly formed groups and coalitions challenged governments and industry over a wide range of pollution and development issues. The growth in environmental consciousness was triggered in part by the publication of works such as Rachel Carson's *Silent Spring*.'[12]

The history of the environmentalist groups closely paralleled the history of the DOE itself. They went through a period of growth in the late 1960s and early 1970s, a decline in influence after 1974–5, a late-1980s period of massive membership expansion, and a shift in focus towards national and global environmental problems in addition to local and regional ones.[13] Environmentalists played a major role in encouraging the Trudeau government to create the DOE, and drew their early strength from 1960s demands for political, social, and economic reforms. In Canada, we find environmentalism thriving within many prominent ENGOs, such as Greenpeace Canada, Friends of the Earth, Pollution Probe, and Energy Probe.

Ecologism and preservationism are more recent bodies of thought. At the core of both of these philosophies is a belief that non-human life has its own intrinsic value and a right to remain undisturbed by man. As Toner notes, the argument is made that man should 'move away from our philosophical roots in anthropocentric reasoning towards a new biocentric philosophical foundation ... humans have no right to reduce the richness and diversity of nature except to satisfy vital needs.'[14] These views have dominated in a few small groups such as Earth First, and have also influenced the conduct of other environmentalist and conservationist ENGOs. Preservationism originated as a radical faction within conservationism but recently evolved on its own towards favouring no-growth solutions rather than the sustainable-use and best-technology solutions supported by conservationists.[15]

Ecologism is also still in its infancy. It affirms, to an even greater extent than preservationism, the need for a holistic approach that testifies to the seamless web between humanity and the natural world.[16] It also holds little faith in the environmental intentions of government or industry, and supports a complete transformation of society. The roots of ecologism can be found in academic discussion over a century or more among biologists. However, it took a more concerted political form in the 1960s and again in the 1980s, and borrowed from modern versions of socialism which advocated radical decentralization. Writings within 'deep ecology' have developed these ideas most fully.[17] Because such views present a fundamental challenge to the status quo, their advocates are seen as the most radical and irresponsible by established interests.

The key point about these philosophies is not that there is a direct fit between a particular philosophy and any one ENGO. Rather it is that a range of philosophies exists across and within groups that greatly contributes to fragmentation but simultaneously supplies the great energy level that ENGOs bring to political action. This range also fragments because it must be channelled into any one or more of the tactical options which the ENGOs employ to influence the policy process either indirectly or directly. The indirect tactics range from simply 'providing concerned citizens with a vehicle for activism' to 'taking extreme positions which broaden the scope of the possible.'[18] The direct tactics also cover a broad scope, from taking part in stakeholder consultation processes to doing research, to actually implementing government policies.

Another manifestation of ENGO fragmentation and energy can be found in the growth and proliferation of groups. One of the earliest DOE studies of the ENGOs showed that, in 1975, there were about 300 ENGOs, 29 of which were considered to be major.[19] By the late 1980s there were 280 groups in the Alberta Environmental Network alone, up from 80 in the mid-1980s, and there were 1,200 groups in Western Canada.[20] The Canadian Environmental Network's (CEN) database lists more than 2,000 groups, compared with only 80 recognized groups when the CEN was formed in 1978.

Throughout the 1970s, the DOE's political and bureaucratic leadership certainly knew that it needed the ENGOs as a supportive constituency, but it was simultaneously reluctant to embrace them. First, there were so many groups that it was difficult to know to whom one should speak. Second, the ideological spread of the groups made it dangerous within the government to be seen speaking to some of them. Third, each of the constituent services within the DOE saw itself dealing with different groups and different kinds of ENGOs. Accordingly, the DOE approached the ENGOs very gingerly in the 1970s, seeking to keep them involved but at the margins.

This pattern of DOE ambivalence about the ENGOs was felt intensely by many early ENGO activists. As one ENGO pioneer put it, 'We felt that [the] DOE had no sense of an environmental movement. They appeared to want limited contact with us and we were broke and felt quite gloomy and alone.' Another early participant likened the DOE to an amnesiac and stated that the 'DOE never had an institutional memory about the groups. Every time there was a new team at the helm we would have to educate them all over again.' One important reason for the apparent bouts of memory loss was an inherent realization in the DOE that if the ENGOs raised too many expectations the department could not ultimately deliver the goods. The ENGOs would therefore need to be managed on the margins of the environmental-policy process, or maybe just ignored.

The early marginalization of the ENGOs by the DOE can be seen in four developments. The first, to be looked at more closely in the next section, is that the ENGOs in the 1970s were basically shut out of the regulatory consultation processes under the Fisheries Act. The second example of marginalization was the environmental-assessment process itself. Once established and confined to projects only, the assessment process served in part to isolate ENGO participation into one corner, albeit an important corner of the environmental world. In arguing this point of view, we are not suggesting that the DOE did not launch the project-assessment process out of genuine environmental concern. Rather, what is suggested is that such a reform had, for a time, the effect of placating the environmental lobby.

But while the ENGOs were often marginalized, changes did gradually occur. As chapter 9, on environmental assessment, shows, the ENGOs used the environmental-assessment arena to full tactical advantage. Gradual changes occurred in the 1970s that broadened the representation and independence of the panels used in such assessments, and the ENGOs, through their pressure and representation, began to affect the substance of the decisions reached by these panels. As mentioned in earlier chapters, it was also in the context of some of these hearings that DOE officials would often unofficially aid the ENGOs by supplying them with better technical data than they could obtain themselves. It must be remembered in this context that another of the consequences of ENGO lobbying activities was that they had both inherited and built up a network of DOE officials who were extremely sympathetic to the cause and who tried to help where they could.

The third example of marginalization, but which simultaneously was a step towards the institutionalization of the ENGOs within the DOE orbit, was the formation in 1972 of the Canadian Environmental Advisory Council (CEAC). Composed initially of leading environmentalists such as Donald Chant, and other ecologists and experts, the CEAC was in part a DOE response to the problem of which ENGOs it ought to deal with. The CEAC was not formed of ENGOs per se, but rather symbolized a kind of beachhead into, and link with, the emerging and growing ENGO community. The CEAC directly advised the minister and was seen by deputy ministers and the minister's staff as a more stable and permanent way of finding out what the environmental community thought.

It was on the recommendation of the CEAC in 1980 that the DOE further formalized its relationship with the ENGOs when it agreed to hold annual meetings with a national steering group. This led to the ENGOs' establishment, in 1982, of an initial national operating structure. In 1987 this structure was incorporated as the Canadian Environmental Network.

The ENGOs, meanwhile, were by no means content to adopt a one-channel approach to access and influence.[21] Indeed, many regarded the CEAC as a co-opting agency. But the issue of being potentially co-opted went far beyond the issue of the CEAC. Indeed, it went to the core of the free-rider problem. The ENGOs needed the state to help them finance their activities. When the DOE developed its first overall policy for funding the ENGOs in 1975, it was already a modest funder of six groups. These included the Canadian Nature Federation, a traditional ENGO, and extended to a group in British Columbia involved in fighting the Skagit Valley project.[22] The BC group included a young Vancouver lawyer, John Fraser, who four years later was to become Canada's environment minister.

The DOE was interested in increasing its funding of ENGOs, and funding reached an annual average of about $150,000 in the late 1980s. But the process of devising a policy for such funding almost immediately involved a search for appropriate umbrella bodies. The DOE was not eager to hand out the money directly among the many pleaders. The problem of finding a central 'peak association' was not to be totally solved, nor could it ever be. But these early funding-policy concerns, coupled with even later more complex policy-consultation exercises in the 1980s, eventually did lead to the formation of the Canadian Environmental Network. Thus, not for the first time did the federal government help establish its own particular national-interest-group structure to assist it in forging relationships with an otherwise unwieldy and diffuse constituency.[23]

The DOE's funding policy for the ENGOs eventually evolved to a point where the fundamental rules of the game were fairly clear cut. Groups that qualified had to show they were largely voluntary, give evidence of having competence to do good work, and demonstrate that they were not beholden to any one private interest in terms of the bulk of their funding. None the less, the funding was always controversial and often had to be augmented by letting contracts to the groups for specific work, frequently under the guise of the department's communication budget. Such funding was frequently the object of attack at budget-cutting time when the DOE's own programs were under siege. Special problems occurred in the early 1980s, when the DOE funded the Coalition on Acid Rain, whose activities focused not only on Canadian lobbying but also, on an unprecedented, visible scale, on lobbying in Washington.[24] An account of this particular episode is best left to chapter 7, on acid rain.

As the 1970s ended, there was a symbolic juxtaposition of the essential journey that the ENGOs had followed in the 1970s. One of their own, environmentalist lawyer John Fraser, a genuine ENGO graduate, was in the minister's chair. During Fraser's period in office, many ENGOs enjoyed their best and most congenial access to the DOE ever. Fraser worked the old ENGO network that he himself had been a part of. But alas, the bubble burst when the Clark government was defeated and the ENGO in the chair departed office. 'So close and yet so far' was not an inappropriate epitaph for the decade then ending. In the 1980s, the marginalizing of the ENGOs was gradually reduced, but to appreciate these subsequent events, we first need to follow the parallel story on the business side of the DOE–interest group equation.

Business and the DOE: From Inner Table to Round Table

While we have already warned against the practice of treating business as a uniform entity, there is one sense in which generalizing across the business–DOE relationship strikes a useful contrast with the points noted about the ENGOs. The first is that business interests in general tend not to suffer from the free-rider problem to the extent that the ENGOs do. While there certainly are inequalities of power among different business groups, they are as a whole more cohesive in philosophy and organization than the ENGOs. The business-interest groups and key firms that interact with the DOE, especially regarding regulation by the Environmental Protection Service, are characterized by considerable cohesion and an intense vested interest in the survival and profitability of their own firms.[25] Moreover, where they are regionally concentrated and powerful industries, such as pulp and paper, asbestos, or oil and gas, they enjoy the additional political clout that is supplied by sympathetic provincial governments. In addition, as we saw in chapter 3, these industries have friends in other Ottawa-based departments as well.

A second reality in the business–DOE relationship which can never be overcome, only moderated by degree, is the ultimate dependence that any social regulator such as the DOE faces in dealing with the widely diverging structures and circumstances of the industries being regulated. This was reflected in the early industrial task-force approach used by EPS in developing regulations. The task forces consisted only of industrial experts. No ENGOs were invited. Task-force members were usually nominated by the main industrial interest-group associations.

The dependence on the industry for the necessary levels of knowledge about production technologies, product characteristics, and environmental effects is overwhelming and is not ultimately attributable to the superior lobbying power of industry per se.[26] Such power is certainly a part of the story, but the dependence is ultimately broader and more inherent in the nature of the practical task at hand. First, regulations must be general enough to span the entire industry or sector involved. This can range from the situation in the pulp-and-paper sector where almost 150 mills are involved to the more managable but still difficult petroleum-refining sector where 26 refineries operate. Second, regulations must accommodate the different circumstances of old or existing plants versus new plants. Third, they must deal with different permutations and combinations of basic production technology being used at different plants. Each member firm in a sector or interest group is always at its own particular stage of launching new investments or seeking to recoup the returns of past investments. World prices and competition affect their views of what is feasible at any given time.

The tiny band of experts in the EPS (at best, no more than two or three per sector) who know these industries are very knowledgable people but ultimately cannot hope to have or be able to replicate or challenge the knowledge that resides in the very heart of the firms which constitute the industry. Accordingly, Canada has relied, as has every other country seeking to regulate firms in the environmental field, upon more general practices agreed upon after consultation with the industry concerned. Most environmental regulation is technology-based and initially applied to new plants rather than requiring a lot of retrofitting of old ones. Such regulation is premised on the idea of encouraging and requiring the 'best practical technology' or 'best feasible technology' for new plants. Old mills or plants were often unregulated, or at least very laxly supervised. For new plants, the theory is that only after all technologically feasible steps are taken will any environmental damage be accepted. Initial compliance monitoring establishes whether the proper technology is in place. A 'monitoring for discharge standard' is developed and determines whether the technology continues to work and whether emissions remain within limits. Regulatory issues are different for water and air pollutants, and varied across the six industries originally picked by the EPS for regulatory action. There is little doubt that this 'end-of-pipe,' technology-based approach has helped reduce emissions, but following an initial burst of regulation in the period from 1971 to 1974, the EPS was unable to progess much farther.[27]

Regulatory initiatives, for example, were also taken regarding the metal-mining, lead-smelting, petroleum-refining, and vinyl-chloride industries. Progress in these sectors varied considerably, but in each case the main processes of regulation making were dominated by industrial involvement. And the main burst of progress occurred early on rather than in the latter half of the DOE's first two decades.[28] A further telling example of industrial power occurred in the context of the forging and implementation of the Environmental Contaminants Act. This 1975 legislation was essentially weakened through the intense lobbying of the Canadian chemical industry. The legislation required the pre-testing of new chemicals, but the law had virtually no teeth and was given little administrative support within the DOE. Continuing dissatisfaction with this law in the wake of growing problems with toxic substances later led to the passage of the Canadian Environmental Protection Act in 1987.

The resistance to further progress in the rest of the 1970s and well into the 1980s came from the industries themselves but was also caught up in the previously examined federal-provincial battle over the Fisheries Act. As mentioned, from the mid-1970s on, federal regulators, especially at the compliance end of the regulatory continuum, were under clear instructions to defer to the provinces, many of which were sympathetic to the actual or claimed economic duress stated by the industries concerned.

It is also even more vital to remember the other federal policies or economic circumstances which were impacting on industry, the national finances, and regional prosperity just at the time that a 'second generation' of environmental regulations was being contemplated in the mid-

1970s. From 1975 to 1978, wage-and-price controls were in place to bring down double-digit inflation. From 1979 to 1981, the second energy crisis hit home with a doubling of oil prices. Then, in 1982, the worst recession since the 1930s added a further blow. It was not difficult under these circumstances for various industries to argue that they should not be hit with new environmental regulatory demands. Indeed, business as a whole in Canada was arguing as the 1970s ended that the regulatory burden of government had to be lifted through regulatory reform. By this they meant primarily a lessening of costly social, including environmental, regulation.[29]

While it is certainly possible to argue that the DOE suffered from a heightened form of industry capture in the 1970s, it was not a uniform phenomenon. For example, the detergent companies mounted a fierce lobby against regulations on the use of phosphates in the early 1970s, but this was resisted by the DOE, in part because of the strength of the overall Great Lakes effort, and in part because the industry was not central to any one province, including Ontario. Later, in the early and mid-1980s, the resistance of the four main SO_2 polluters, including INCO and Ontario Hydro, was eventually overcome, in part because Canada had to show its own good intentions through action in the acid-rain battle with the United States.

In speaking of the DOE's relationship with business, not to mention that with the ENGOs, it is also important to differentiate the business clientele of other DOE services from that of the EPS. Parks Canada, for example, learned fairly early on that it had little choice but to build public-consultation processes into its Parks policy, both in general and with respect to the establishment of new parks. As one senior Parks Canada official put it, 'We did it as a practical matter of self-preservation.' This consultation approach also became necessary because, in the realm of parks, even the business side of the constituency could be quite diffuse. Typically there were many more small businesses involved in and around the parks. Moreover, the parks sector, as we see in chapter 8, had to deal with an aggressive tourism lobby.

Indeed, typically, Parks Canada officials would find themselves surrounded by three sets of interest-group clusters: very strong-willed environmentalists who literally wanted to protect the park from all human use; a local group of businesses, local workers and citizens who basically saw the park as a source of jobs; and usually some middle group who saw value in both preserving and using park lands. Parks Canada invariably had to seek consensus somewhere close to the third group, and hence it found its local public hearing-and-meeting structure to be essential in this process.

Another area where business certainly did not always get its way was in the realm of environmental assessments. While many approached the minister, the deputy minister, or other Ottawa politicians to call off the dogs of environmental assessment, once an assessment was under way the DOE did not succumb to such pressures. It must be remembered, of course, that a lot of the discretion as to whether a hearing had to be held did reside with other line departments. Gradually, however, in the 1980s, the assessments became more and more stringent, and hence many firms had to learn to live with, if not love, them.

Business–ENGO Convergence: New Maturity or New Battles?

It is difficult to pinpoint when some sense of potential convergence between business and the ENGOs and between both and the DOE began to occur, or exactly how to characterize it. Perhaps it was symbolized in 1986 by the composition of Tom MacMillan's ministerial office. It contained a vigilant and skilled ENGO kindred spirit in the person of Elizabeth May, but also experienced industrial advisers such as Ron Wosnow, a former Imperial Oil official. May became a key player on issues such as the South Moresby Park decision examined in chapter 8 and persuaded MacMillan to offer the ENGOs a commitment to the park's establishment in his very first public speech as a minister. Indeed, MacMillan had invited May to join his ministerial staff on the explicit understanding that she would be his ambassador to the ENGOs and their voice in his office. But the steps that led to this semblance of greater balance in the business–ENGO equation begin much earlier.

In 1980, Blair Seaborn convened separate meetings between himself, selected ENGO representatives, and business leaders. In each case about eight to ten persons were invited to exchange views. Many business and ENGO personnel certainly faced one another at environmental hearings, but these were arenas for combat, not mutual discussion. Another catalyst was the process used in 1981 to forge a policy on public consultation announced later in October 1981. Other than having a policy on funding the ENGOs and a consultation policy for Parks Canada, the DOE as a whole had never established a comprehensive policy on consultation. As we have seen, the *de facto* policy at the EPS was that only industry was at the table, through its industry-task-force approach. The 1981 policy was followed by a fairly elaborate exercise in 1982 in which the DOE held consultation meetings across the country.

The impetus for better consultation came from many sources in the DOE but, as always, it came even more forcefully from the ENGOs themselves. None the less, John Fraser's strong interest was taken up by his successor, John Roberts. An early advocate at the senior level of the DOE was Jacques Gerin, especially in his capacity as Senior ADM Policy. The acid-rain issue was also an important catalyst. The DOE could hardly back the Coalition on Acid Rain openly and publicly and then imply that public consultation was not important for other areas as well.

ENGO–departmental relations were, however, by no means on a serene upward path. Following the stabilization of ENGO funding in the first half of the 1980s, the ENGOs encountered problems when Genevieve Ste Marie became deputy minister in 1986. She cut the support which the ENGOs had been receiving through various direct and indirect routes. After protests that the cuts were unfair, the deputy replaced the varied kinds of support with a straight fund of $150,000. This amount was the DOE's estimate based on what the previous forms of support had totalled, but the ENGOs felt it was a considerable shortfall. When Elizabeth May became MacMillan's ENGO ambassador, she was charged with dispensing the new, diminished resources. Her tactic, in contrast to previous practices, was to hand it out in very small amounts to the numerous, small, scattered ENGOs rather than to the larger, somewhat better-endowed ones.

However, the particular event that many point to as the beginning of a new convergence, or closing of the gap, between business and the ENGOs was the 1984 Niagara process. The Niagara Institute is a private organization devoted to fostering improved contact and discussion among the key institutions in Canada, particularly business, government, and labour. The forum for such discussions is its comfortable, old Ontario resort headquarters at Niagara-on-the-Lake. The Niagara Institute had fostered its own mode of bringing interests together, including the use of conferences and third-party facilitators and intermediaries.

The manner in which the Niagara process eventually led to the DOE's patenting of its own multi-stakeholder process was more accidental than planned. When Charles Caccia became minister in 1983, he handed his officials a list of about twenty-five items he wanted launched or looked at. One of these was a conference on jobs and the environment. The original DOE official who was to coordinate the process was unable to continue and, as a last-minute substitute, the Niagara Institute was asked to organize and run the exercise. Initially business interests were unenthusiastic about any such event, especially because they feared that Caccia was planning a big, splashy, media-style conference rather than one in which real progress could be made. When business groups said they would not be a party to such a road show, the search was on for exactly how to hold a constructive multi-stakeholder event.[30] Eventually it was decided that the first Niagara event would be a discussion on 'plant modernization,' a euphemism for environment, economy, and jobs.

In January 1985, an initial meeting of about fifty persons drawn from business, the ENGOs, labour, and the federal and provincial governments ground to a halt. The old habit of firing shots across one another's bows took hold. Yet there was enough concern among some of the participants that they were able to spend some time examining why genuine consultation was not occurring. As Glen Toner points out, the postmortem concluded that 'there was a serious problem with language, participants from different groups simply did not speak the same language, they did not have the same understanding of the problem, nor did they have the basis to trust each other.'[31]

As a result, it was decided to give the exercise another try. Four working groups were set up and given four months to prepare reports. One group was given the task of developing a 'Niagara Manifesto' which would reaffirm the willingness to work together. The second group worked on developing a common language and terminology. At a later plenary session in June 1985, the reports of these two groups converged to produce what some described as a kind of etiquette for consultation, a list of key points that had to be achieved in order to have a positive consultative atmosphere.

The third group worked on the issue of environmental data so as to give the various interests something closer to a common information base. It eventually agreed that such data would have to be independent of, or at arm's length from, government if it was to have credibility among the interests. The work of this group was a further impetus to the DOE's adoption of State of the Environment reports.

The fourth group was given a problem rather than a general issue to deal with – namely toxic substances. But without the benefit of the first three groups, which were meeting concurrently, the toxics group floundered. The toxics group also had a more difficult problem. It had to tackle an actual regulatory and policy issue. Here the old instincts took hold again, in part because the DOE itself displayed too much missionary zeal in seeking to persuade other participants of the virtue of its 'cradle to grave' approach to the regulation of toxics. Despite this, the group decided to work at it for a few more months. Again, no consensus was found.

At this point, a new catalyst concentrated the stakeholder minds. This came in the form of a strong indication from the minister, Tom Mac-Millan, that he wanted legislation on toxics. The toxic exercise was no longer an abstract one, and this served to energize the group, which, though its members had not agreed, were certainly getting to know one another personally. For the new legislation, MacMillan secured Cabinet approval that would allow him to have the stakeholder groups work directly on draft legislation, allowing them to see exact legislative terminology and wording. After several extensive meetings and negotiations, the process yielded the draft statute which eventually became the CEPA, and which enshrined the life-cycle, cradle-to-grave approach to the management of toxic chemicals.

The Niagara process seemed to break the ice and allow new forms of contact and interaction. It was during and after this process that other influences seem to converge as well in an uncoordinated but interactive way. The articulation and popularization of the sustainable-development paradigm supplied the umbrella under which diverse interests could seek comfort from the environmental rain. Inside the DOE, these various efforts and influences were increasingly referred to simply as the 'environment–economy thrust.'[32] Another outgrowth of the new synthesis was the formation, following the visit to Canada of the Brundtland Commission, of the National Task Force on Environment and Economy. Many of the same Niagara players took part in the task force, along with many others. They showed up again in the work of the Energy Options process that the energy minister had set up to advise him on Canadian energy-policy. Indeed, for the ENGOs in particular, this was by far the greatest involvement they had ever had in energy policy matters.

Another example of the new multi-stakeholder mode of consultation and consensus formation came in the process used to develop a twenty-year management plan for NOx and VOCs emissions in Canada.[33] The ENGOs were primarily represented through the Canadian Environmental Network. Business interests covered a range of sectors far more complex and numerous than had been involved in the acid rain and SO_2 management process. The Canadian Council of Ministers of the Environment coordinated the process, and the DOE supplied the technical staffing and secretariat. Independent facilitators were hired to run the consultation sessions. In short, the Niagara process was in full swing.

A further key influence on the dynamics and content of business–ENGO interaction also emerged from a development already referred to in previous chapters – namely, that business interests in the late 1980s and early 1990s had a much more complex take on matters environmental and were being pressured more directly, both by public opinion in the wake of the global-warming scare of 1988 and by other financial institutions which were now insisting that corporations put their environmental houses in order. By the early 1990s, the business community was, at a minimum, a set of four different interests regarding environmental matters. First, it contained a growing number of firms which make their living from selling environmental equipment and technology and from retailing products that are environmentally friendly.[34] These interests have already created a momentum of their own that can only grow and with whom the DOE can build more supportive alliances. Second, there are several large sectors, such as telecommunications and computers, which see themselves as clean industries and which wish to differentiate themselves publicly and competitively from the dirty or environmentally irresponsible industries.[35] For example, Bell Canada developed its first explicit corporate environmental plan in 1989. Some of the impetus for the Bell Canada plan came from the company's participation in the CEPA consultation process in 1986–7. Other parts of the impetus came from shareholders and employees concerned about environmental issues and about future bottom-line profitability.

Third, there are the traditional polluting industries. Many of these, such as the chemical and pulp-and-paper industries, were rushing to develop both the perception and the reality of a better environmental record. For example, the Canadian Chemical Producers Association adopted its Responsible Care Program of the late 1980s in part as a reaction to the Bhopal, India, chemical-plant disaster of 1985, but also in response to wider political pressures. The Canadian Pulp and Paper Association also sought to improve its environmental practices and established its first industry-wide environmental charter.

Fourth, and perhaps most decisive of all, other components of the corporate world, from auditors to lenders, to securities regulators, were increasingly involved in monitoring companies as to their current and future environmental liabilities. These pressures have begun to hit the corporate bottom line in ways that will vastly exceed the impact of general CEPA-style regulation.

Also emerging from the same learning experiences were the many environmental round tables that were formed to advise the federal government and the provinces and various private-sector groups about how to institutionalize sustainable development.[36] Indeed, by the early 1990s, business and the ENGOs were sitting together in so many places that a group of them decided to meet bilaterally, without the presence of the DOE or provincial mediators. Calling themselves the New Directions Group, corporate executives from companies such as Dow Chemical and Noranda and representatives from ENGOs such as Pollution Probe and the Canadian Nature Federation met intensively during 1991.[37] One of their first areas of agreement was a policy position, presented to the federal environment minister, on toxic substances, including an agreed-upon schedule of implementation for eliminating several high-priority toxic hazards. Such a bilateral move would have been literally unthinkable even three years earlier.

While there is little doubt that some convergence of interests has occurred and, with it, a more mature approach to the environment, it is also appropriate not to underestimate the potential for future environmental conflict between the ENGOs and business, and between both and government. Several factors make this likely.

First, the ENGOs are still financially and technically vulnerable, and yet will be expected to be, and will want to be, involved in ever-more-complex regulatory and policy processes, both domestic and international. Hence new conflicts over funding are bound to occur. Second, the use of the courts and the demand for an environmental bill of rights are likely to increase. The *Rafferty-Alameda* and *Oldman River* cases of 1990–1 are unique in Canadian environmental history but must be seen against a larger backdrop of developments in the 1980s which have seen Canadian politics being driven by a rights-oriented approach.[38] The influence, symbolically and practically, of the Charter of Rights and Freedoms on environmental strategies by business and the ENGOs should not be underemphasized. And last, but not least, the environmental movement still contains groups with a very diverse range of environmental philosophies and tactics, all of which will undoubtedly be employed in the 1990s.

Public Opinion: The Villain Is Us?

At the end of the 1970s, political scientist Michael Whittington accurately summed up the DOE's biggest failing when he said that the DOE had not succeeded in creating or taking advantage of a supportive constituency.[39] The DOE deserves some sympathy in this regard, not only because of the pressures on it inside the Ottawa system but also because mobilizing public opinion of such a general nature carries with it some of the same free-rider problems encountered by the ENGOs. The constituency of opinion is so broad that it is hard to articulate and focus it beyond a few basic generalities, such as wanting a clean environment and getting tough with polluters. Similar problems are faced by the consumer movement, the women's movement, and other broadly based public-interest constituencies. Canadians are all to some extent simultaneously voters, producers, workers, consumers, polluters, and family members residing in regions of a vast country. It is far easier to mobilize Canadians on some of these attributes than on others.

None the less, the DOE tried to build on what it saw as a steadily increasing base of supportive public opinion, if not for the DOE, then certainly for environmental issues writ large. When one looks at public opinion and media coverage over the twenty-year period since 1970 it is true that the underlying base of support did gradually broaden. But within this overall increase there were clearly smaller peaks and valleys that were extremely important.

The best evidence of the early burst of strong public support came in the early years of the 1970s, in the form of federal and provincial throne speeches and legislative commitments that launched the environmental age. Thereafter public interest in the relative priority that should be given to environmental issues plummeted, as shown by their sparse representation in front-page media coverage, and the decline of political interest mirrored it in the content of Throne speeches and the relative absence of debate on those issues in Question Period. Only the burst of increased concern that followed an environmental accident would disturb this picture until the early 1980s.

For example, the quantitative assessment of media coverage which the DOE obtained regularly showed in 1979 that environmental coverage had moved from last place to tenth place on a list of about twenty-five issues.[40] Much of this improvement was accounted for, even in the midst of the huge energy-policy crisis of 1979, by the fact that the acid-rain issue had been publicly and graphically linked to the existence of hundreds of 'dead lakes.' The DOE's own polls showed that environmental issues in 1979 were viewed by Canadians as being more important than national unity and that Canadians believed that environmental goals could and should be achievable without a loss of economic well-being.

By 1983, the trend assessors were telling the DOE that environmental issues were continuing to hold their own, at about 5 per cent of media content coverage. Toxic chemicals were a new evident concern, and the reports showed that there was growing impatience with regulatory laxity, especially the lack of willingness to prosecute polluters.[41] Whereas previously the DOE's polling activity had been intermittent, by the mid-1980s, it was a regular feature, carried out through three separate polling and survey companies. The polling was also increasingly subtle. Polls in the summer of 1986 told the department, on the one hand, that the public was generally quite satisfied with the quality of the environment, but, on the other hand, that it was increasingly worried about toxic substances.[42] The safety of nuclear power had also moved up the scale of concerns because of the recent Chernobyl reactor accident. The data also revealed that the vast majority of Canadians were more likely to believe and trust information provided by environmental groups than they would information from either business or government. And 60 per cent believed that governments were doing too little to regulate industrial pollution.

By 1987, media coverage had ratcheted up yet another notch, to 7 per cent of coverage, and the DOE's pollsters were telling the department that Canadians were looking for long-term solutions and a multiple-issue approach in recognition of the greater complexity of problems. Polls were also showing that Canadians saw themselves as part of the problem, with increasingly high percentages saying that they had made lifestyle changes out of concern for the environment.

By the summer of 1989, the upward march continued, and with it the DOE's feeling that its day in the political sun had finally arrived. Environment was the first 'top of mind' issue, cited as the most important problem by 16 per cent of Canadians, ahead of the deficit, unemployment, and free trade.[43] The polls also showed that Canadians in all regions looked first to the federal government to deal with environmental issues.

The Green Plan, the ENGOs, and Business

It was at the height of the public-opinion surge that Ottawa launched the 1990 Green Plan. Alas, public opinion does not design green plans; ministers, officials, and interest groups do. Previous chapters have shown the role of the former in the Green Plan policy process but not the role of ENGOs and business.

In June 1989, the Canadian environmental community presented the prime minister, Brian Mulroney, and the environment minister, Lucien Bouchard, with their sustainable-development strategy, called *A Greenprint for Canada*. It was crafted by a coalition of more than forty environmental and Native organizations. While it had a line of argument and an internal logic consistent with the values and positions of these interests, it clearly involved both a radical agenda and a green 'wish list' that a Conservative government would not fully endorse. The *Greenprint* agenda also alarmed much of the business community. It did, however, add to the pressure on the government to create its own version of a sustainable-development strategy. The Cabinet was also increasingly listening to the ENGOs' and business criticism that there was not enough real consultation built into the Green Plan–formulation process.

When the draft DOE Green Plan was taken back to Cabinet for approval in January 1990, the Cabinet did not decide yes or no on the plan but instructed Bouchard to develop a process to consult with the public. The consultation process that followed between April and August 1990 was extensive – information sessions in thirty cities, two-day multi-stakeholder workshops in seventeen cities, and a final session in Ottawa in August. However, the consultations satisfied very few, in part because everyone knew they were an 'add-on' to the process, since the original budget secrecy–related time limits imposed by Minister Bouchard simply did not allow for consultations. Thus, many stakeholder representatives were already in a bad mood as the consultations began, and some were even more upset when they discovered that the document developed for the consultation was not the actual policy document taken to Cabinet in January, but a discussion paper which asked a series of questions and did not contain any financial details.[44] This was hardly surprising since Cabinet had not yet decided on a universal dollar figure for the program as a whole, nor had they approved any details of the plan. Still several thousand Canadians became involved in one phase of the consultation or another.

As a result of these dynamics, the ENGOs were extremely suspicious, and business was increasingly edgy. The lack of specific provisions in the content of the discussion document meant that consultation sessions were not directed to either choosing priorities or exploring detailed programmatic options. Many charged that the real document was locked in a desk back in Ottawa, while the consultations were just window-dressing. The only organization to benefit explicitly from the consultations was the DOE. The organization and management of the consultations across the country brought the historically diverse department together in an unprecedented way. But, to top matters off, Lucien Bouchard resigned as minister of the environment, and from the government, just as the consultation process was beginning. It was left up to cabinet veteran Robert de Cotret to guide the Green Plan through its final stages.

The immediate ENGO and media response to the Green Plan was highly critical. A major criticism was that the final plan contained no specific program details, insufficient costing data, no legislation ready for immediate tabling, and no firm federal-provincial agreements. Environmentalists had argued for a radical policy employing green taxes, a 'big stick' approach to regulation and emission-control standards which would have moved Canada out in front of other OECD countries. They were almost unanimously critical of the Green Plan, decrying its 'cautious tone' and charging it with being too strong on intended federal-provincial agreements, new research, and public education, and too weak on tighter laws and regulations, and stricter enforcement. The plan was criticized for not including an environmental bill of rights, for not creating an independent environmental auditor to oversee federal policy, and for delaying the imposition of green taxes and other financial incentives for further study. Environmentalists did approve of some parts of the parks, wilderness, and wildlife sections of the plan, while others approved of the government's intention to undertake a comprehensive review of the environmental implications of all existing federal statutes, policies, programs, and regulations.

In the main, the federal Green Plan has not been viewed very favourably by Canadian ENGOs, in part because of those groups' ingrained habit of applying absolutist criteria. Disappointment is voiced at the absence of any immediate commitment to greenhouse taxes. The failure to adopt all, rather than simply most, of the ideas generated by the public consultation process is viewed, not as normal democratic practice, but as an index of failed intentions.

Such tactics are, of course, practised to some extent by all interest groups, but they are much more endemic to the environmental lobby because, in many respects, components of the environmental lobby do seek a radical restructuring of society. The radical component practices the studied politics of entrenched impatience. To its credit, the environmental community's continuous pressure is in part why the federal Green Plan has as much in it as it has. But this raises a further tactical issue which environmental lobbyists too often forget. If they have a deserved stake in some of the content of the existing Green Plan, then they may very well have a stake in giving the government some visible non-partisan support for its initiative. As we have seen, the DOE has failed to develop a supportive external pro-environment constituency. It may well be in the 1990s that the shoe will be on the other foot. If the environmental lobby is unrealistically critical of the Green Plan and other initiatives, it will weaken its *only* beachhead *within* the confines of the government. As argued in chapter 3, important constituent parts of the bureaucracy remain hostile to efforts to undertake major environmental-policy planning.

As for business interests, they were reasonably pleased that the Green Plan avoided immediate new regulations and taxes and committed the government to public and industry consultations before developing further regulations in areas such as controlling greenhouse-gas emissions. Business liked the promise that the Green Plan will be reviewed on an annual basis, arguing that this is the sort of 'self-adjusting plan' approach that business would take itself. Some business representatives did charge, however, that the plan would create more bureaucracy and impose additional costs on industry.

For business, however, the most serious problem with the Green Plan is that it fails to demonstrate a serious commitment to the use of market-based policy instruments as a complement to traditional regulation. There is the obligatory mention of such devices as tradable pollution permits and environmental taxes but a failure to show why they are essential if there is to be any hope of achieving the scale of change needed, and of achieving such change at optimum social cost in an increasingly competitive world. It is the relationship between traditional regulation and market incentives that is the key, but federal environmental-policy makers buried these issues in the deeper recesses of the report with a promise to provide an additional discussion paper, so as not to offend environmentalists and some business interests as well.

Many business groups privately believe that the Green Plan is far too expensive, given the state of the Canadian economy, but they are keeping their public criticisms to the bare minimum. Some of this studied tactical silence is undoubtedly attributable to their own preoccupations with the greening of their own companies. As we saw above, in the the late 1980s and early 1990s many Canadian firms did begin a serious review of their own environmental practices. Some of this effort was propelled by the same general public pressure that governments were feeling. But much of it was also a result of the even more telling interest of banks, insurance companies, securities regulators, and auditors about the current and future environmental liabilities to which a company might be exposed.

There is no doubt, then, that business interest in the Green Plan was significant – some of it supportive, but much of it concerned with slowing down the DOE policy juggernaut that was moving through the Ottawa policy system. As examined in chapter 4, Western Canadian oil and gas interests in particular, fearful of a carbon tax, portrayed the DOE bureaucracy's aggressiveness as a replay of the interventionist 1980 National Energy Program.

The absence of any environment taxes in the Green Plan can undoubtedly be attributed to business pressure, but such tax measures were unlikely to have been looked upon favourably by the federal Department of Finance in any event. This lack of enthusiasm was owing in part to analytical uncertainty as to just what kind of environmental taxes to impose but even more to the concurrent imposition of the Goods and Services Tax (GST), at the forceful instigation of the Department of Finance. While polls showed that Canadians were tolerant of environmental taxes if the funds were used for environmental purposes, the finance department was convinced that voters would punish any government that gave them both the GST *and*, if environmental taxes, and both in the midst of an economic recession. But if business was in a defensive posture during the Green Plan–formulation period, it is likely to go on the offensive during the much longer Green Plan multi-year implementation phase.

Conclusions

The temptation to stereotype the DOE's relationships with business, the ENGOs, and the public is often overwhelming. According to this view, business sees the ENGOs as naïve and perhaps dangerous 'no-growth' addicts, and it sees the DOE as being only a milder reflection of the same anti-economic bias. The ENGOs see business as a capitalist environmental villain who still has the DOE in its hip pocket to such an extent that environmental laws have at best only periodic barks and certainly no lasting bite. Public opinion is seen as generally supportive of stronger environmental measures, but, alas, the public sees others as polluters, but not themselves. Moreover, in the public mind, both business and government are not to be trusted on environmental matters, but especially the former. Business still seems content to ride out each blip in the opinion polls following major accidents or spills, knowing that something is changing but hoping that it is just another cycle in public opinion.

But while this stereotyping has some validity in understanding the early years, this chapter has shown that it loses its economy of description in coming to grips with the 1980s and early 1990s. And in the final analysis, it is not good enough to understand fully the earlier period either. An understanding of these three elements of the DOE's constituency requires a realistic 'bottom-up' mapping of just what it means to speak of 'business,' 'ENGOs,' or the opinions of the public in environmental matters. For example, the business clientele of the Environmental Protection Service consisted of industries such as pulp and paper and mining, but Parks Canada dealt more with local Chambers of Commerce and numerous hotel operators. Equally, an ENGO for the EPS might mean a national body such as Pollution Probe, while for the Wildlife Service it may be Native groups concerned about fishing and hunting rights. For the Parks Service, the ENGOs may be local conservationists and numerous kinds of parks users. Similarly, the structure of public opinion varies considerably when questions are posed about the public's views on the environment in general, as opposed to its views as the users of parks or in relation to threats to local jobs. Thus the structure and dynamics of the DOE's interest-group clientele have some of the same fluidity and complexity that we stressed in our previous discussion, in chapter 4, of the imperatives of federalism, area, and power.

The chapter has also shown that, by the early 1990s, the ENGOs and business were functioning in a somewhat more mature relationship. Indeed, they were increasingly inclined to want to deal with each other and sometimes, at least, to cut out the DOE as an interlocuter. The DOE's capacity to deal with and mobilize a supportive constituency had improved greatly in the 1980s, in part through its own efforts, in part through the tide of events, and in part through a slow begrudging recognition by all the interests that the environmental villain was often 'us' rather than always 'them.' But the Green Plan process also showed that there were still significant areas of conflict and profound disagreement.

6

International Environmental
Relations

The final institutional relationship in the federal environmental-policy process is that of international environmental relations. This involved a large and growing array of direct bilateral political and policy dynamics with the United States and with international institutions and numerous other countries on the multilateral front. The Department of the Environment's capacity to exercise influence on the Canadian, continental, and international environmental agenda was also conditioned by the inherently more constrained nature of foreign policy and international relations. Unlike domestic-policy making, where largely all policy instruments (spending, regulation, taxation, and persuasion) can be employed, foreign-policy making must rely more on persuasion and diplomacy achieved through protracted negotiations. International relations also involved diverse relations within the government between the DOE, the Department of External Affairs (DEA), and the Prime Minister's Office (PMO).

The international realm, however, was also an arena for sometimes symbolic and sometimes real environmental leadership. Indeed, for many of the key players in the DOE's early days, the department was often visualized as mainly a domestic reflection of the ambitious international environmental values and programs that the DOE espoused on Canada's behalf from the outset of the environmental age. From the Stockholm Conference of 1972 and the World Conservation Strategy of 1980 to the Brundtland Commission of the mid-1980s and the Rio Earth Summit of 1992, there has been a strong Canadian influence on the international environmental stage. The careers of international environmentalists such as Maurice Strong, Jim MacNeill, and David Munro span these events in which Canadians articulated a broad and generous global environmental agenda. Indeed, the paradox of this disproportionate influence was that, while ambition and vision could be shown internationally, they could not as readily be delivered at home. Thus international meetings became a kind of welcome respite from the environmental trench warfare on the home front.

International environmental relations are examined in this chapter in six sections. First, the Great Lakes Agreement is analysed as the quintessential example of 1970s–style Canada–United States cooperation. Second, we look more generally at the United States as both hero and villain on the Canadian environmental stage. The next two sections deal with the evolving multilateral arena of environmental policy, initially by linking the array of international institutions that have emerged to the strategies involved in forging the '30 Per Cent Club' in the 1980s negotiations on sulphur-dioxide emissions, and then by showing the DOE's new dilemmas in complex protocol-setting processes in the latter half of the decade. These include the difficulties experienced over the Protocol on Oxides of Nitrogen (NOx) and the relative success and leadership shown in the Montreal Protocol on CFCs. The fifth part of the chapter explores the DOE's role in helping to foster international environmental-policy paradigms to advance the overall environmental cause, most importantly that of sustainable development. Finally, the chapter looks at Canada's relations with the Clinton Administration and the extension of environmental foreign policy into the trade-policy realms of the General Agreement on Tariffs and Trade (GATT) and the North American Free Trade Agreement (NAFTA).

As the international environmental-policy journey is analysed, it is helpful to keep in mind three features of the two-decade period as a whole. First, bold environmental internationalism could initially be practised mainly in the multilateral arena of Canada's foreign-policy relations. There, big, broadly progressive ideas could at least be discussed, very much in the tradition of Pearsonian-style foreign-policy making.[1] Things were very different in the bilateral continental world of Canada–United States environmental relations. Big ideas and environmental grand designs simply did not impress the Americans, especially in the Reagan era of the 1980s. The bilateral arena, whether on issues as large as acid rain or as small as a local river-pollution issue, was characterized by dogged day-to-day relations, where the test, as often as not, was whether a sufficient coalition of U.S. interests could be brokered, badgered, or bought. Coalitions, of course, had to be brokered on the Canadian side of the border too, but the U.S. congressional system, with its separation of powers between executive and legislature, functioning in a fifty-state federal system, made Canada more often a policy taker than a policy maker.[2]

Second, as the global commons came to be more fully understood over the two decades, it was increasingly impossible to keep the bilateral and multilateral realms separate.[3] The processes and dynamics of reaching international agreements and setting protocols became ever more complex and changed the relations within the government between the DOE and the DEA, as it did among the larger array of domestic and international interests involved. Successive negotiations on SO_2, the ozone layer and CFCs, NOx, VOCs, global warming, and biodiversity involved hazards that made international issues increasingly complex.[4] With the election in 1992 of the Clinton Administration, a new chapter was opened that increasingly embraced trade and economic issues, not the least of which was NAFTA and the battle over its environmental side agreement.

A third key to the analytical journey is that environmental internationalism has involved successive efforts to propagate new paradigms of environmental thinking such as the concept of sustainable development. These processes have thus produced a significant level of learning by osmosis through persistent political, economic, professional, and scientific contact.[5] International relations have therefore not just been state to state but also scientist to scientist, business to business, and ENGO to ENGO, as an international policy community evolved and matured, lock-step with, and often ahead of, the patterns we have already seen domestically.

Bilateral Relations and the Great Lakes Agreement

One central fact underpins environmental management between Canada and its giant neighbour to the south – that a country whose economy and population are ten times larger than its neighbour can do much more environmental damage to its neighbour's than it can to the larger economy. Faced with this reality, the history of environmental relations shows that Canada in general and the DOE in particular have had to rely on four interwoven strategies. First, they have urged and relied upon a principled response, much of this initially anchored in the work of the International Joint Commission (IJC). Second, they have sought to show that U.S. actions can damage other Americans, an approach which requires the building of national or local alliances on both sides of the forty-ninth parallel. These first two strategies are best exemplified in the promulgation of the Great Lakes Water Quality Agreement of 1972. Third, they have alternately elevated the United States to the position of hero and declaimed it as villain on the environmental stage, often adopting its best practices and frequently deriding its worst excesses, for both domestic political and practical advantage. And fourth, they have sought to 'out-science' the United States wherever possible.

The adoption of a principled approach by both countries is best seen in relation to the pre-DOE concepts entrenched in the Boundary Waters Treaty of 1909. In this treaty, Canada and the United States agreed to three important elements that have since defined the relationship not only in relation to water pollution but also to a considerable extent in matters of air pollution.[6] The first element was the principle that one nation may not take certain actions unilaterally within its borders which could injure the other nation. The second was that Article IV of the treaty provided for an outright prohibition against injurious transboundary water pollution. The third was the establishment of the International Joint Commission.

The treaty applied primarily to water flows, but there was a brief but strong reference to water quality as well. The definition of injury was left to determination in each case, and hence the structure and operating norms of the IJC became pivotal. The IJC was composed of six members, three from each country. Its procedures were intended to avoid the conduct of business along national lines and instead were focused on solving problems based on the facts of a situation as seen at least several steps removed from the direct clash of interests. Thus a tradition of collegiality, independence, and objectivity developed as decisions worked their way through the IJC's technical advisory boards, which were themselves mini–technical commissions with co-chairpersons and an equal number of Canadians and Americans. On this basis, the IJC could be asked to approve applications for raising the level of waters; to enquire and report into matters raised by either or both countries; and to arbitrate between the two parties when requested by the two parties.

When the DOE was formed, the IJC continued to function in its important work, and many of its principles and operating habits rubbed off on the bilateral wing of the DOE's international role. The DOE, mainly through Inland Water, but also through the Environmental Protection Service (EPS) and the Atmospheric Environment Service (AES), provided most of the technical expertise for the IJC boards.

The Great Lakes Water Quality Agreement, however, was forged fully in concert with the IJC model. A brief account of the political economy of this agreement demonstrates not only its benchmark importance in international relations and its dominant effect on the early agenda of the DOE, but also how things changed under the constellation of forces in the acid-rain equation in the 1980s. The Great Lakes Agreement was being concluded in the DOE's first year of existence by an External Affairs and DOE team which included External Affairs official Ray Robinson, who was later to hold senior DOE positions in Canada–United States Relations, the EPS, and the Federal Environmental Assessment and Review Office (FEARO); Jim Bruce; and Alan Prince. Though signed in 1972, the agreement was preceded by an earlier reference to the IJC in which both governments asked the commission in 1964 to report on Great Lakes water quality. A cooperative scientific effort led to a major report in 1970, to which DOE staff made a major contribution and which helped pave the way for the later agreement.

Indeed, as we have already seen in our brief account in chapter 4 of the Canada Inland Waters program, one of Canada's early strategies in dealing with the United States was to 'out-science' the Americans on Great Lakes pollution. This tactic was deemed necessary in order to provide a convincing description of the nature and extent of the pollution damage. While this approach was successful in some respects, particularly regarding the effects of phosphates and their substitutes, it was not in itself a sufficient condition for political success. Equally important was the need to forge a coalition of interests on the U.S. side of the border. This meant primarily a coalition among the several populous states that bordered the Great Lakes, such as New York, Michigan, Ohio, and Illinois. These states were the source of the pollution but also stood to benefit from the clean-up. Lobbied hard by local U.S. environmentalists, these states faced the main political problem of levering money out of Washington's coffers to pay for the needed sewage-treatment plants.

Both Canadian and U.S. environmental interests engaged in mutual back-scratching and lobbying, usually in a quiet, behind-the-scenes manner. Little of the loud public diplomacy so evident in the acid-rain case was needed or contemplated. Canada possessed modes of access to Washington as a foreign government that state governments could not match. On the other hand, Canada needed the lobbying clout and potential votes of populous states to be successful in congressional politics. Moreover, in sharp contrast to the later 1980s, there was money to be had in the Washington and Ottawa public purses of the early 1970s.

The Great Lakes Agreement was not without controversy but, compared with the acid rain saga, was reached relatively quickly and amicably, as were the amendments to it in 1978 and 1987. It established specific water-quality objectives and mechanisms for improving them further. It respected each country's different ways of reaching these objectives. It also provided for a Great Lakes Water Quality Board and a Great Lakes Science Advisory Board, and included various mechanisms under which the IJC could publicly report, at its discretion, progress on the agreement. Finally, it specified that both parties had obligations to seek the appropriation of funds to implement the agreement, a special concern on the U.S. side among the Great Lakes–state governments. A complementary Canada–Ontario agreement was also signed to ensure that the Canadian side could deliver on its commitments under the international agreement.

There is little doubt that the Great Lakes Agreement was the quintessential example of constructive Canada–United States cooperation. It was also the environmental success story that the DOE most used in the 1970s to demonstrate its relevance and efficacy. Indeed, in many ways it was deservedly the focal point for an internal sense of pride within the DOE. The Great Lakes project was in many respects the engine of the DOE, but one which propelled the department in different directions. Pride was clearly felt by those who were a part of it, but there was also some resentment in other quarters in the department at the resources and attention obtained for the Great Lakes initiative. None the less, a number of other important boundary-water issues were solved by the IJC–DOE cooperative effort, including the Garrison Diversion project in North Dakota–Manitoba, and the Skagit issue in British Columbia. Inland Waters still maintains a steady watching and monitoring brief on more than forty boundary-water areas.

The United States as Environmental Hero and Villain

The importance of highly visible projects like the Great Lakes initiative cannot be overestimated. But they each require a sense of perspective. For example, the acid-rain case during much of the 1980s was a symbol of both environmental frustration for the DOE and bad environmental relations with the United States. Thus the United States went from being a comparatively cooperative player in the Great Lakes case to villain in the acid-rain story. This is not to suggest that there is not considerable accuracy in both these labels, but in each decade they also cloud, for the public, a great deal of normal day-to-day environmental-problem solving carried out over a friendly border. They also cloud other examples of Canada's inherent ambivalence about things American.

What then did normal bilateral relations look like? During the 1970s and 1980s, there were approximately 100 issues and controversies active in the Canada–United States file. These ranged from the Great Lakes Water Quality Agreement and acid-rain developments to a host of smaller issues such as local agricultural run-off problems and pulp-and-paper industry pollution on particular rivers.[7] In any given year, however, eight or nine issues could be considered to be major concerns, many, of course, with a life cycle of several years. Thus problems such as the Garrison Diversion or Lake Erie pollution might persist. Many of these problems might not be large in national terms but could be quite central in regional terms. Thus, almost invariably, middle-sized international issues were simultaneously federal-provincial issues, with local ENGOs and businesses applying intense local pressure.

In this setting, the Bilateral Relations unit of the DOE and its barely half-dozen officers functioned as a combined negotiating nerve centre and post office for the ongoing problems that emerged. Working with, and keeping an eye on, the IJC, the bilateral unit had a threefold role: to select the more contentious issues from among the many that were identified by the DOE's scientific monitoring; to advise and decide (through consultation with the key technical experts in the line services) on what to do about them; and to seek solutions to problems at the lowest possible level in the bureaucracies of both countries. The rule of thumb was that the higher up the ladder decision making went, the more difficult it would be to find solutions. It is in this sense that the DOE sought to emulate the norms of the IJC, albeit without the arm's-length structure. Thus the most normal and frequent form of contact and problem solving was not through the two countries' embassies but rather through extensive day-to-day telephone and personal contact between DOE officials and their counterparts at the Environmental Protection Agency and in other line departments in both countries. Each line agency of the DOE had a U.S. technical counterpart, and probably 90 per cent of transboundary issues were solved or dealt with at this level.

There is no doubt that this system has resulted in the solving of problems and in the successful management of many agenda items, any one of which, if not properly dealt with, had the potential not only to explode in a DOE minister's face but also to do environmental damage. The agenda included major issues such as the Garrison Diversion, where a U.S. project prejudicial to Canada was vastly scaled down after both an

IJC study and persistent DOE lobbying. As we see later, the acid-rain case is noteworthy for the manner and extent to which it violated the norms and habits forged in the IJC and early DOE tradition.[8]

There is a second reason why there is a strong need to ensure that the environmental relationship with the United States is not interpreted only on the grand scale of 'heroes and villains.' This reason is that, as is the case in Canada's larger political relations with its superpower friend, Canadian perceptions are characterized by an ingrained ambivalence about U.S. influences. For example, early environmental reformers in fields such as environmental assessment looked with admiration at the openness and tough requirements of U.S. legal requirements. As chapter 9 shows, they wanted an equally tough statutory base for Canadian environmental assessments. Instead, Canada adopted a guidelines-based approach because other political and bureaucractic influentials feared a kind of U.S.-contagion effect. They became obsessed with the idea that Canadian environmental processes could become tied up in endless U.S.-style litigation.

A second example of Canadians' ambivalence can be found in attitudes to U.S. openness regarding information and data. Canada's ENGOs, on the one hand, still get some of their best information, technical and otherwise, under U.S. freedom-of-information provisions and from the vast U.S. scientific enterprise. Yet, in many other scientific and technical controversies, they will castigate Canadian regulators if they adopt U.S. standards without doing their own retesting of data.

And finally, consider the brief environmental flurry that occurred during the great free-trade debate in 1987–8. In this case, a national coalition of ENGOs attempted to portray the United States as a grossly inferior environmental country whose lower environmental standards would seep into Canada if free trade was adopted. This charge occurred despite the fact that in many areas – such as auto emissions – U.S. standards were far tougher than Canadian ones and hence no such across-the-board labels could even remotely be considered to be an honest description of reality.[9]

Much of this later view of U.S. environmental villainy was a function of considerable Canadian opposition to the practices and attitudes of the Reagan Administration, particularly between 1981 and 1985. The Reagan Administration was none too subtle in its efforts to weaken the role of the U.S. Environmental Protection Agency (EPA) and to stonewall international environmental progress.[10] We examine this retrenchment in greater detail in the acid rain case study in chapter 7. Suffice it to say that for the twenty-year period as a whole, the DOE's bilateral relationships are probably best seen as characterized by a duller and more business-like normality rather than by either the warm glow of the Great Lakes Agreement or the depths of the acid-rain Reagan-era depression, important though both those events undoubtedly were.

Multilateral Relations: Towards the '30 Per Cent Club'

The multilateral realm of environmental relations was one in which the DOE often felt instinctively more comfortable than it did in the bilateral arena. Indeed, to some extent in the 1970s, the bilateral and multilateral worlds could be kept distinct. The bilateral world required talk and action, while the multilateral world was, in the 1970s, somewhat more in the realm of talk and research. In the 1980s the bilateral and multilateral arenas became increasingly entangled in concert with the changing nature of the environmental problems being faced and in light of the tactical need to build alliances and to solve mutual problems.

In this section we focus on the evolution of one organization, the Organization for Economic Cooperation and Development (OECD), and one episode, the forging of the '30 Per Cent Club' in 1984 to give some of the flavour of the evolution of the DOE's multilateral role. But first, a basic map of the international organizational terrain is obligatory. Six organizations in addition to the OECD have some role in international environmental matters, each with different memberships and powers, and hence each involving different degrees of difficulty in alliance building and consensus formation.[11] In general, the larger the membership of the body in question, the more difficult the dispute-resolution processes and possibilities.

The United Nations has three organizations with environmental roles. The U.N. Environmental Program (UNEP) is the largest and most important.[12] It grew out of the Stockholm Conference and was initially headed by Canada's Maurice Strong. Its purpose is to promote international environmental cooperation, to review the world environmental situation, and to report on the implementation of environmental programs. The United Nations' Economic Commission for Europe (ECE), a smaller body than UNEP, composed of both Western and Eastern European countries (including Canada and the United States), became more important to Canada in the early 1980s. Its mandate is broadly economic in nature, but it became a key forum on issues such as the Convention on the Long-Range Transport of Atmospheric Pollutants. The U.N. Educational, Scientific and Cultural Organization (UNESCO) is the third U.N. arm but the least active in environmental terms as a whole. It does, however, promote world heritage and other aspects of man and the biosphere in which DOE officials have been active.

Three other international agencies of considerable importance are the World Meteorological Organization (WMO), the International Maritime Organization (IMO), and the International Union for Conservation of Nature and Natural Resources (IUCN). The WMO promotes world meteorological cooperation, information, and research, and as such is a key part of the operating context for the DOE's Atmospheric Environmental Service (AES). Indeed, both the AES and the WMO became even more important lead players as more research was needed in the 1980s on the complex issues of global warming and climate change. The IMO is engaged in promoting practicable standards in marine safety and navigation, including pollution from ships. Accordingly, it is a key element in implementing agreements such as the 1973 International Convention for the Prevention of Pollution from Ships and the 1978 Tanker Safety and Pollution Prevention Convention. And last but not least, the IUCN is primarily a non-governmental body which promotes international cooperation in applying ecological concepts to the conservation and management of nature and natural resources. As shown further below, it was an important player in advancing the concept of sustainable development and in promoting international paradigms for environmental thinking.

In the many international meetings fostered by these organizations there emerged a DOE and a Canadian role, a *modus operandi* that was quite different from the bilateral role, especially in the 1970s and early 1980s. Individual DOE officers, usually scientific and technical persons such as Fred Roots, Jim Brydon, David Munro, Peter Bird, and John Hollins, were in a position to, in effect, 'speak for Canada' and for the DOE, often without a detailed authority or mandate. In the distant locations of these frequent meetings, such officers would be able, in effect, to commit Canada and the department to environmental 'high road' positions, at least verbally and in terms of the commitment of ideas. As a result, Canada became known for progressive environmental advocacy even where there was a gap, as there frequently was, between what could be said abroad and what could be delivered at home.

This did not mean that there was no contact between the DOE and the DEA on these issues. Instead, these situations arose because the DOE, all other things being equal, was more preoccupied with bilateral and domestic issues, leaving, in effect, a vacuum in the multilateral arena. Moreover, the international meetings themselves produced a real stimulus, or contagion effect, on the scientists who attended them. Just being there and learning about problems created an enthusiasm for ideas and solutions that, at home, would be sucked dry by the inherent density of bureaucratic layers and jurisdictions. In truth then, key DOE officials frequently became international-environmental policy entrepreneurs who increasingly used international meetings as a way to keep the pressure for reform on the players back home. This phenomenon is best seen in a closer look at the one international organization we have yet to profile, the Organization for Economic Cooperation and Development (OECD), and in the episode concerning the '30 Per Cent Club,' a part of the politics of acid rain involving the ECE as well.

The OECD is primarily an economic organization of the wealthy Western economies, but among its dozen working committees there has been an active Environment Committee. This committee is nominally directed by the member countries' environment ministers, but they met initially only about every five years. Therefore the committee was almost always dominated by officials. Its concerns in the 1970s and early 1980s were threefold in nature. First, it was concerned about harmonizing the basic environmental requirements of OECD countries in the interests of promoting fair competition. Second, it did important early work on issues such as the regulation of chemicals and the shipment of hazardous wastes. Third, because of its large European membership, it had an early preoccupation with the use of coal. While each of these was an important activity, the coal connection is of special interest to us in seeing links to the later establishment of the '30 Per Cent Club.' Europe had ample coal resources and was anxious to expand their use in light of the two oil-energy crises, in 1973 and 1979. But the problem with coal was that it was perceived to be, and is, a 'dirty fuel.' At this stage 'dirty' referred to its sooty characteristics – in short, visible dirt – rather than to the later concerns about its contributions to acid rain and global warming. The OECD Environment Committee's subcommittee on energy and the environment was especially exercised by the concern for enhancing the use and image of coal as an alternative fuel.

In the early 1980s, however, attention began to shift from these narrowly focused concerns to broader issues. Rather than just the effects of coal on the environment, attention came to be focused on the environment as an input to resource use. Some of this shift in focus can be attributed to Canadian influence. During this period, Jim MacNeill headed the OECD Directorate on the Environment in Paris and his often-expressed more holistic approach to environmental issues began to make its presence felt. But Canadians in the OECD network were not the only ones advocating a change in view. The Scandinavian countries were also involved in research on issues such as acid rain and began pressing for reform.

It was in this sense that the international forum contagion effect began to take hold. The contagion was fuelled in part by growing concern about what the research on acid rain was showing but there was also a second vital element to the contagion phenomenon – namely, that both the Canadians and the Scandinavians had suitable environmental villains, the United States for the former, and the United Kingdom for the latter. These countries were the main sources of acid-rain pollution for Canada and the Scandinavian countries respectively. But, even more important in political terms, neither Reaganite America nor Thatcherite Britain was known in the first half of the 1980s for its environmental vigour. In the early meetings, the West Germans often behaved much like the Americans and British but this situation changed after research showed adverse effects on German forests and after the Green Party gained sudden prominence in West German politics.

But even this international advocacy and alliance building among officials and scientists would not have borne fruit had there not been allies at home. These came in the form of a change in tactics in dealing with the Americans and involved a trio of DOE players, the minister, Charles Caccia; the director general of Intergovermental Affairs, Danielle Wetherup; and the ADM of the Atmospheric Environment Service of the DOE, Jim Bruce. Each in his or her own way reached the same conclusion early in 1984 – namely, that bilateral tactics with the Americans were going nowhere and that a multilateral card had to be played.

Wetherup and Bruce had accompanied Caccia to one of his regular Washington meetings on the Garrison Diversion controversy early in 1984. After a particularly depressing meeting with the Reaganite environmentalists, Wetherup and Bruce both urged Caccia to take the lead in the formation of what Wetherup dubbed the '30 Per Cent Club,' a group of countries that Canada would mobilize to announce publicly their commitment to reduce SO_2 emissions by 30 per cent by 1993. The entire purpose of the exercise was to exert pressure on the Americans and the British. On the plane home from Washington, the three devised the basic shape of the multilateral strategy.

At this point, however, the issue also involved the choice of the appropriate international forum. The ECE became the forum of choice, in part because it contained members from Soviet Eastern Bloc countries, countries which both produced and were adversely affected by acid rain.

Each member of the Canadian '30 per cent' trio played his or her necessary role. Caccia held a highly public meeting of the 'club' on 22 March 1984. At the meeting, Canada and nine European countries signed a declaration of their intentions to reduce SO_2 emissions by at least 30 per cent by 1993. This target was not a problem for Canada because it had already committed itself to even larger reductions, of 50 per cent, by 1994. For her part, Wetherup worked the key part of her bureaucratic network, paying special attention to the key ECE delegates from Sweden and Norway, sarcastically labelled by one American delegate as the 'Canadian–Nordic boy scout trio.' Jim Bruce meanwhile was appointed chairperson of the ECE working group which negotiated the Helsinki Protocol. From this base he also was building on his contacts with the Russians, who chaired the ECE, and with key Russian scientists.

The '30 Per Cent Club' was eventually expanded to twenty-two countries, who signed an ECE-sponsored protocol in July 1985. The United States and the United Kingdom declined to sign the protocol. Accordingly, the strategy employed to use the mutilateral card can be judged only as, at best, a modest success. Indeed, other officials in the DOE regarded it as only a political sideshow. After all, it was another six years before the Americans took action under amendments to the U.S. Clean Air Act. The episode is also instructive in conveying a sense of the extent to which DOE officials could function as veritable environmental Lone Rangers during this period. It also reveals one of the first occasions where the bilateral and multilateral worlds were joined tactically and substantively, a situation that would increasingly be the norm in the 1980s and beyond.

But the episode of the '30 Per Cent Club' was also, in many respects, a unique configuration of events that allowed the DOE's symbolic environmental leadership to occur. These limits were acknowledged by a report on international relations prepared in 1989 by the DOE's International Affairs Branch. It observed that the relationship between domestic and international environmental agendas was changing.[13] In the early 1980s, it observed, Canada's domestic policies and practices 'exceeded our international obligations' and that this 'enabled Canada to play a leading role in the development of agreements such as the Helsinki Protocol on SO_2.'[14] The report stressed that, in the case of SO_2, international multilateral considerations were 'not crucial in domestic program development.'[15]

In short, the DOE could afford to take the international high road because it had done its technical homework and had taken strong action at home. But, as we see below, other hazards requiring multilateral negotiation would not find Canada or the DOE to be as certain of its scientific base, nor would key domestic interests be as nicely aligned like ducks in a pond as they were by 1983–4 in the '30 Per Cent Club' episode.[16]

These changing international dynamics also influenced the relations the DOE had to develop with the DEA. With respect to bilateral affairs, the DEA much preferred the quiet diplomacy of executive-to-executive relations. But, at the same time, given the DEA's much larger overall foreign-policy responsibilities, it deferred to the DOE on by far the largest part of the numerous transboundary issues. Only on the acid-rain case were there some tensions as to the DOE's preference for a noisier form of diplomacy and agitation, but even here, External Affairs was willing eventually to take the gloves off and use the bare-knuckle approach as well.

On the multilateral side, as we have seen, both the DOE and the DEA initially held the reins much more loosely. But this approach began to change in the latter part of the 1980s as the protocol-setting processes became ever more complex, and interrelated with other economic and foreign-policy matters.

The New Political Economy of Protocol Setting

As the DOE went through its learning curve of successive international agreements and protocols, on SO_2, lead in gasoline, and CFCs, the new political economy of negotiating protocols became increasingly evident to the DOE's international team in the Corporate Policy Branch. But they could not always act on these new understandings, nor did outside stakeholders see the new dynamics clearly. Indeed, what was beginning to collide was the increased complexity of international processes, on the one hand, and the equally complex relationships now developing with the DOE's ENGO–business stakeholder consultative processes. A brief look at the early dynamics of negotiating agreements and protocols on nitrogen oxides (NOx) and volatile organic compounds (VOCs) is instructive in this regard.

As stressed above, in the mid-1980s, Canada's international credentials on the environment were very good. Its level of preparation on SO_2 negotiations was excellent. The situation was different at the beginning of the NOx–VOCs story.[17] In September 1986, West Germany, Switzerland, Austria and several other countries – under domestic pressure from their respective Green Parties – began to develop standards on NOx emissions, especially on autos, that were already weaker than those of Canada. Many of these countries had not had a good environmental image abroad, and new evidence showed that their forests were deteriorating badly, in part owing to their own NOx emissions. Along with the Nordic countries, these states pressured Canada and other countries to develop tougher standards on NOx emissions: 30 per cent reductions were called for.

This was one '30 Per Cent Club' that Canada decided, for good reasons, it did not want to join. The reasons were not hard to find. First, the European countries with the worst records were seeking a 30 per cent reduction from a high emission base. It made for good environmental rhetoric in the home political market. For countries that had already been better environmentally, further 30 per cent cuts were far more difficult and might not make sense relative to taking action on other pollutants. But it was difficult to manage the media politics involved in saying that a country was opposed to such targets. DOE spokespersons who had to defend Canada's position likened it to a fat person dieting to lose 30 per cent compared with a thin person facing a 30 per cent weight loss.

Further media spice was added because sitting in the meeting rooms at the NOx negotiations were representatives of environmental groups, one of which issued a report critical of Canada. Although rarely interested in such usually uneventful meetings, the media quickly picked up this assessment and brought additional pressure to bear when Ontario's minister of the environment agreed that Canada should join this new '30 Per Cent Club.'

There was also greater scientific uncertainty in the NOx case than there had been in the case of SO_2. DOE scientists did not know to what extent NOx was a serious problem for Canada's forests. Knowing the impact of extreme NOx emissions, as were prevalent in Europe, was a scientifically different and less difficult problem than knowing the impact of low NOx emissions, the situation in Canada. DOE scientists did know that, in the Canadian context, NOx was also important, in combination with VOCs, in producing urban smog. But this problem varied widely across Canada. It was clear again, therefore, that it was not going to be in Canada's interests to agree to across-the-board percentage cuts.

Percentage cuts, moreover, were part of a larger set of genuine policy controversies about how to deal with hazards. Control measures could range from the adoption of ceilings on emissions to percentage reductions, to the adoption of best-available technologies and the use of critical loads. In general, the DOE was advocating the use of critical loads.

These were the dynamics in 1988. On 31 October, under the aegis of the ECE, Canada signed the NOx protocol, which requires it to freeze NOx emissions at their 1987 level by 1994 and to implement further actions – which must be negotiated with the United States – to reduce them to the level required to achieve internationally agreed-upon environmental-quality targets. Meanwhile, the ECE had also established a working group to develop recommendations for a VOCs-emissions-control protocol.

Earlier, the Canadian Council of Ministers of the Environment (CCME) and the DOE had come under fire for the lack of domestic consultation regarding negotiations with the United States over acid rain. Now the energy industry, through fierce lobbying directed at the External Affairs department, roundly criticized the DOE and the NOx protocol because Canada had made stronger commitments than energy interests could tolerate. They also argued that they had not been properly consulted. Moreover, it was known that NOx–VOCs emissions would be a key part of the revisions to the U.S. Clean Air Act then being negotiated in Congress. These pressures led to the CCME's decision in October 1988 to launch an elaborate NOx–VOCs consultation process, which would lead to the development of a comprehensive plan to manage and reduce emissions.

Thus the NOx negotiation process exhibited problems for the DOE and was in many ways viewed as a way not to reach protocol agreements. It was often contrasted with the forging of the earlier Montreal Protocol on Substances that Deplete the Ozone Layer. That protocol was viewed as being exemplary, though hardly easy in the making. Canada is not a major producer of chloro-fluorocarbons and thus could approach the growing ozone-layer-depletion hazard with considerable objectivity and leadership.[18]

The Montreal Protocol was adopted in 1987 and provided for an initial set of targets for reducing the production and consumption of ozone-depleting chemicals. It also set up a detailed process for periodic scientific monitoring, which led to further changes to the targets in 1990. The overall Montreal Protocol program included targets that would take into account the availability of substitutes, environmental impacts, and the economics of different control options. It was also one of the first to recognize the legitimate claims of Third World countries for funding to finance their adaptation to greener environmental products. The DOE's role, especially through key officials such as Vic Buxton, was instrumental throughout, but it was especially important in suggesting models and approaches for handling the varied interwoven technical and economic aspects of control options.[19]

While the above scarcely does justice to the full story of the NOx–VOCs process, let alone the Montreal Protocol decisions, suffice it to say that these cases as a whole typify the main features of the new political economy of international environmental-protocol setting.[20] These features include the existence of technically complex, multiple interwoven hazards; the presence of a large array of domestic industries and interest groups affected by any action or strategy; the involvement of dozens of countries with their own array of coalitions to accommodate; and the inevitable presence of intense media and ENGO pressure.

All of these attributes were present in even greater measure as the issue of global warming and the ozone layer came on the agenda in the mid-1980s and early 1990s.[21] Global warming was dealt with earlier, in our discussion of the role of the Atmospheric Environment Service in chapter 1. But several features of its effect as an issue on the general nature of international environmental relations are especially important.

The first effect is that the massive international media and public attention paid to the issue, combined with the ending of the Cold War, has elevated environmental issues to the point where they are now being considered as matters of international security.[22] Thus, they are already becoming a set of issues that will transform the nature of all international relations. As such, it is argued by many that global 'ecopolitics' presents a 'fundamental challenge to our traditional realist-based, state-centred models of international relations.'[23]

The second effect, and a reason that traditional international relations are likely to be so challenged, is that the international environmental movement, as a political lobby, is now unambiguously global. The 1992 Rio Earth Summit showed this starkly, but it is in fact manifested in several ways, including the emergence and initial electoral success of Green Parties in many countries; the alliances among domestic ENGOs to draw world media attention to green issues such as through the issuance of green report cards to the leading G-7 countries at Economic Summit meetings; and the linkage of environmental values and issues as a moral challenge to the established world order.[24]

The exact shape of a new international order in which environmental issues become seen as security issues cannot yet be fully described. But some would agree with Lynton Caldwell when he concludes that the evolving international environmental movement 'belongs to a larger transformation in human social thought, which may be likened to a second Copernican revolution.'[25] In Caldwell's view, 'the first revolution removed the earth from the centre of the universe: the second removes humanity from the centre of the biosphere.'[26]

From Stockholm to Rio: Environmental Paradigms and International Agendas

While the bilateral and multilateral worlds were tended to in the fashion outlined above, there is another layer of international dynamics that warrants separate treatment. This layer came in the form of four key international conferences and events whose purposes were to articulate new or evolving environmental-policy paradigms and to advance the global environmental agenda. By 'paradigms,' we mean more or less coherent philosophies and ways of doing things within which a policy field gains coherence and momentum. In each such event a distinct Canadian presence helped to energize these efforts to keep the environmental ball rolling in the face of powerful counterpressures. The four events are the U.N. Conference on Human Settlements held in Stockholm June 1972, the 1980 conference that produced a World Conservation Strategy under the auspices of the International Union for the Conservation of Nature, the work and report of the World Commission on Environment and Economy (the Brundtland Commission), and the 1992 Rio Earth Summit.[27]

The Canadian and DOE involvement in these paradigm-forming exercises and, in turn, their influence on the DOE are an important part of the international and domestic environmental story. The Stockholm Conference, chaired by Canada's Maurice Strong, was the first international effort to draw worldwide attention to environmental programs and to suggest an agenda and workplan for their resolution. Chapter 2 has already shown the extent to which the Stockholm ethos influenced the DOE's thinking. It was at Stockholm that many of Canada's first environmental policies were really articulated as a conceptual plan. The early DOE internal planning documents constantly referred to it and tried to build on it.

The effort to devise a World Conservation Strategy in 1980 was less obviously successful in the advancement of environmental ideas and practices than was Stockholm. But it was attempted in 1980 by the IUCN in part out of growing frustration that the nations of the world were giving short shrift to their environmental good intentions, and in part out of early concerns, long before they were politically fashionable, about issues such as ozone depletion, and even global warming. The World Conservation Strategy recognized the key linkages between nature, conservation, and development, and articulated the concept of a sustainable utilization of resources. At the same time, it sought to avoid the trap of the 'no-growth' mentality which many first-generation environmentalists had so often displayed.

The DOE involvement in the World Conservation Strategy was spearheaded by Dave Munro. The impact of this effort to advance environmental values and paradigms was muted in Canada and elsewhere, largely because of the energy crises of 1979–80 and then the devastating recession of 1982. But within Canada, and certainly within the DOE, there were significant pockets of opinion which embraced the conservation ethic. Bodies such as the Science Council of Canada had also attempted to sell the concept of a 'conserver society' well before there was an audience willing to listen to the message.

But by 1984 some of this intellectual preparation began to influence the international body politic. There is a direct causal link between the work of the OECD's Environment Committee referred to earlier and the formation and work of the Brundtland Commission. And there is equally another vital Canadian connection in the person of Jim Mac-Neill, who headed the OECD's environmental secretariat, who had been a key player in Stockholm, and who had advocated the more holistic concept of the DOE as early as 1970. MacNeill became the secretary general of the Brundtland Commission. By 1984, the OECD's economic establishment was prepared to say to the world community, which it did at a conference on environment and economics in January 1984, that environment and economic development could be made to reinforce each other, but only if they were integrated at the earliest stages in central decision making in government and industry and at home.

Not surprisingly, when the Brundtland Commission reported, it also concluded under the banner of its environmental paradigm of 'sustainable development' that sustainable forms of development were possible. Brundtland defined sustainable development as 'development that meets the needs of the present without comprising the ability of future generations to meet their own needs.'[28] Canada's environment minister, Charles Caccia, was a participant at the benchmark OECD meeting in 1984 and came home as a genuine supporter of the Brundtland Commission process. The OECD–Brundtland breakthrough worked its way through the Canadian policy system in a variety of ways. The report of the Macdonald Royal Commission on the Canadian economy picked up these central themes through direct DOE pressure.

The DOE, then in the Blais-Grenier doldrums, had eagerly hooked its political star to the new godsend of 'sustainable development' and the economy–environment linkage. The new paradigm was credible enough to purge the DOE of some of its previous image as a flabby, marginal, no-growth–oriented department, but the paradigm was also vague enough to appeal to a wider constituency of support, or at least potential support. The Brundtland Commission toured Canada, and its work led directly to the commissioning of Canada's own Task Force on Environment and Economy. As chapter 5 showed, this task force went a long way towards legitimizing and advancing a new form of relationship between business and the ENGOs and between both and the state.

The final benchmark event was the Rio Earth Summit of June 1992. Unprecedented in its scope of involvement by political leaders, governments, business, and ENGOs in the developed and developing world, the Rio Earth Summit has supplied a further catalyst for change. With its adoption of a massive Agenda 21 program, and its early work on various agreements on climate change, biodiversity, and forestry, there is little doubt that Rio will influence international and domestic agendas throughout the 1990s and beyond 2000.[29]

It is often easy to dismiss these major events and conferences as paradigm and agenda setters. After all, not all the talk leads to action. Such dismissal would be a great mistake, however. The nations involved learn at such events, each in its own way, and each through the osmosis of professional and political contact and education that such meetings foster. Umbrella concepts such as sustainable development (about which much more will be said in later chapters) are needed to foster agreement and to establish momentum even where interests do not agree on the exact meaning of the words. In this sense, environmental internationalism is no different than trade or defence policy, where clarion calls for free trade or security shelter many meanings and many conceptions of national interest.[30]

The Clinton Administration and the Greening of GATT and NAFTA

Some of these competing policy terms and international collisions were given a surprisingly early test in the 1992–3 period and centred around trade issues in general and the Clinton Adminstration and NAFTA in particular.

GATT has been preoccupied since the mid-1980s with the Uruguay Round of trade negotiations and, for these reasons, has not wanted to deal extensively with trade–environment issues. However, the long-dormant GATT Group on Environmental Measures and International Trade had to be revived in 1991 to begin exploring concerns which GATT member countries knew needed some attention and which will be central to the next round of GATT negotiations. The agenda for this group includes trade measures in international agreements and their compatability with GATT rules; the transparency of environmental regulations; and the trade effects of packaging and labelling requirements.

Environment–trade linkages have been forced on GATT both through concerns by developing countries (linked to post-Rio developments, as noted above) and through individual cases such as its own panel decision report on the United States–Mexico tuna and dolphin dispute.[31] The GATT panel had found against unilateral trade measures the United States placed on the import of tuna harvested with purse-seine nets. The nets were killing many dolphins. The panel argued that GATT did not allow a country to impose such unilateral trade measures to apply in such an extra-territorial fashion.

The tuna/dolphins case helped, along with NAFTA, to politicize trade–environment issues, especially in the United States. But it was hardly the first time that such links had occurred or received political notice.[32] First, GATT recognizes the right of countries to develop and implement adequate domestic environmental policies on a non–trade discriminatory basis. Second, policies such as those in the tuna/dolphins case are serious in that a trade measure based on perceived undesirable process and production methods (PPMs) is very problematical. And they are serious because any trade restriction is problematical if its purpose is to protect a resource outside the territory of the country taking the measure. Third, trade measures may be used to induce countries to sign international environmental agreements. This was done, as we saw above, on agreements such as the Montreal Protocol which prohibited trade in certain products with non-parties.[33] The argument in these protocols is that such measures are needed as political and economic leverage and also to ensure that non-signatories do not gain at the expense of those countries which have taken pro-environmental choices.

While negotiated actions on these issues through GATT will likely await the next round of GATT negotiations, the trade–environment issues escalated far more swiftly on the bilateral or continental front. NAFTA was the first major trade agreement to be negotiated in which environmental issues were explicitly involved. There are essentially two international institutional developments involved, one in the actual negotiated agreement, and the other in the 'side deals put forward by the United States that were being negotiated in 1993.

In NAFTA itself, the key environmental provisions include:

- a basic commitment to sustainable development;
- the right of each party to determine the level of protection required to protect its environment and its human, animal and plant life or health;
- the establishment of a work program to enhance levels of protection throughout North America;
- the right to maintain standards that are higher than those recommended by international organizations;
- recognition that NAFTA countries should not lower health, safety or environmental standards ('create pollution havens') to attract investment; and
- provision for special scientific review boards to advise dispute settlement panels if environmental issues are raised.[34]

The last of these provisions is probably the key institutional feature in that it recognizes the far greater importance of scientific input into all environmental matters than in typical trade matters.

While the above provisions within NAFTA are clearly important, the 1993 side-deal provisions are also vital in political-institutional terms. When the Clinton Administration pressed for more guarantees regarding both labour standards and environmental matters and proposed NAFTA commissions on both, the Canadian position on the latter, the initially proposed North American Commission on the Environment (NACE), was to stress its broader facilitative role rather than a regulatory role. Unlike the situation in the Reagan years, Canada's environmental policy was now seemingly on the defensive with regard to the United States. Thus the proposed agency was seen by Canada as having functions such as harmonizing towards higher environmental standards; developing limits for specific pollutants; promoting environmental sciences and technology; collecting data and reporting on the state of the environment; increasing public awareness on environmental issues; and generally cooperating with the overall NAFTA free-trade commission to achieve the overall environmental goals of NAFTA.

Canada strongly opposed any institution's being able to impose trade penalties on those who violate existing environmental standards, as advocated by U.S. negotiators. This position was based in part on the principled view that trade rules should not be expanded in their use for this purpose, but also in part on the specific tactical fear that U.S. interests, environmental and industrial, would have yet another harassment-style trade remedy device, this time fuelled by the tactical and populist agenda of the potent U.S. environmental lobby.[35]

In the final agreement, a Commission for Environmental Co-operation was established, comprising a council, a secretariat, and a public advisory committee.[36] Recommendatory and advisory roles are stressed in the new commission's mandate, including a public-awareness function. On the difficult issues of enforcement, the overall agreement provides different disciplining mechanisms between the United States and Mexico, on the one hand, and Canada and its continental partners, on the other. Where any of the countries fail to enforce their environmental or labour laws and do not correct the problem, they may be subject to fines paid into special environmental and labour funds. In Canada, fines will be enforced by domestic courts. The United States and Mexico, in contrast, will face suspension of NAFTA benefits based on the size of the penalty (in the form of a duty or trade sanction).[37]

The battle over the above institutional arrangements thus raised key political and economic concerns about just how facilitative versus regulatory trade–environmental institutions should be, especially when one of the participants in an agreement is a developing country and is the suspected primary offender in the eyes of the environmental lobbies in the other two.

Conclusions

Canada's and the DOE's international environmental relations have evolved from a period in the 1970s where bilateral relations with the United States could often be effectively partitioned from multilateral relations. In the later 1980s and early 1990s, however, the two worlds of international environmental relations increasingly clashed, in part because the nature of environmental problems had become more complex, and in part because the high-road, big-idea posture taken multilaterally increasingly clashed with U.S. Reagan-style bare-knuckle approach that replaced an earlier bilateral tradition of principled cooperation. More recently, the Canada–United States environmental relationship has become entwined with a Clinton Administration and a U.S.

environmental lobby that have mobilized around trade issues, including both GATT and NAFTA.

The negotiation of international agreements and protocols also became vastly more complex in all of their dimensions: the interactive nature of pollutants; national variations in what the relevant scientific problems and priorities were; the growing range of domestic and international interests involved; and the emergence of numerous international environmental agencies and bureaucracies.

Despite this growing complexity, Canada, through the DOE, played a largely constructive role, not only in particular protocol agreements but also in the four benchmark events that defined the international environmental agenda of the 1970s, the 1980s, and the early 1990s: the Stockholm Conference, the World Conservation Strategy, the Brundtland Commission, and the Rio Earth Summit. By the dawn of the 1990s, global warming and other complex environmental issues were beginning to transform the very nature of international relations, especially since they were being energized by a global environmental political movement of great importance.

7

Acid Rain

No environmental issue better symbolized the dilemmas and dynamics of Canadian environmental-policy making than acid rain. For most of the 1980s, it seemed to be the dominant pollutant on the Department of the Environment's priority list and, for the Canadian public, it was the environmental issue with which voters most easily identified. Unlike some other pollutants, acid rain was quintessentially middle class, because its most visible presence was in Central Canada's middle-income cottage country. The image of dying lakes was one to which many Canadians could relate directly whether as cottage owners or as visitors to the summer home of a friend or relative.

'Acid rain' became the popular name given to what is essentially a complicated environmental hazard.[1] More accurately, it is a form of acid deposition that occurs when emissions of sulphur and nitrogen compounds are transported through the atmosphere. These emissions are then transformed by chemical processes and deposited back again on earth, as either wet or dry depositions. The high level of acidity, which then can adversely effect lakes, fish habitat, trees, and buildings, is a product of all of these factors: the emissions; the nature of the transport; the kind of transformation that occurs; and the deposition, including the specific sensitivities of the receptor areas.

While more will be said later about the scientific and technical dimensions of acid rain, it is important to stress from the outset that the acid-rain story recounted here is essentially that of the battle to control sulphur-dioxide emissions only. Like most of the DOE's major environmental sagas, the acid-rain story is one involving all of the key relationships traced thus far in this book. It involved a pitched battle with the United States, especially during the Reagan era; the need to orchestrate workable coalitions of action with the provinces, especially Ontario and Quebec; the dictates of cajoling and persuading a handful of key firms in the smelting and power-generation sectors to change their production practices; and the need both to utilize and, at times, to deflect the pressure of a determined environmental lobby eager to use the acid-rain case as its beachhead of environmental progress. The dynamics of decision making on acid rain also involve important issues regarding the scientific capacity of the DOE, the politicization of science, and how the dossier was handled by various organizational components of the DOE.

The essence of the acid-rain chronology is portrayed in table 4. For our purposes, the journey begins in the latter half of the 1970s, when the problems of acid rain began to receive more concerted scientific attention, and ends in 1990 with the passage of the U.S. Clean Air Act amendments, actions which finally brought the United States on side. More specifically, we break the acid-rain story into three broad periods, during which the three key relationships – bilateral, federal-provincial, and business–government – changed. These produced, in turn, a short period of hopeful action, a long period of international stalemate but some domestic progress, and then, as the 1990s began, a sense of both renewed hope and action on both sides of the forty-ninth parallel.

Getting and Staying on the Agenda

How a particular environmental hazard gets on the agenda and then stays there is more art than science. It is, in short, ultimately more a political and economic phenomenon than a scientific one.[2] But there is little doubt that science must ring the first credible alarm bells. So it was in the case of acid rain. It began as a part of a more generic concern for the 'long-range transport of airborne pollutants,' or LRTAP, the name preferred by DOE scientists. As a technical acronym it was eventually recognized by the more politically astute that 'LRTAP' just would not make it in the big leagues of political sloganeering. Hence 'acid rain' became the vastly more saleable code-word, aided and abetted by images of dying lakes and environmental time bombs.

But the early association with LRTAP remained important not only because it is technically the case that acid rain is connected to other long-range air pollutants, but also because its early journey through the DOE is associated first with the Atmospheric Environmental Service (AES) and with the Air Pollution Control Directorate, a section of the DOE eventually disbanded in 1982.

TABLE 4
Key Events in Acid-Rain Policy

August 1976: Early scientific concern results in establishment by DOE of integrated program on Long-Range Transport of Airborne Pollutants (LRTAP)

November 1978: U.S. congressional resolution leads to formation of the Canada–United States Research Consultation Group on the Long-Range Transport of Air Pollutants

November 1979: Member countries of the U.N. Economic Commission for Europe sign the international Convention on Long-Range Transboundary Air Pollution. Both Canada and the United States are signatories.

July 1980: Coalition on Acid Rain formed at the urging of John Fraser. Group of environmentalists led by Adele Hurley and Michael Perley takes up the cause, lobbying in the United States and in Canada.

August 1980: Canada and the United States (Carter Administration) sign a Memorandum of Intent which confirms the mutual goal to take concrete cooperative steps to combat acid rain.

April 1981: Governments of Canada, Quebec, Manitoba, and New Brunswick agree to hold acid-deposition loadings to an initial target of less than 20 kg per hectare per year by 1990.

June 1982: Acid Rain Negotiations with the U.S. Reagan Administration under the Memorandum of Intent collapse. U.S. establishes program for more research on acid rain.

February 1983: Reports of Canada–U.S. work groups are released. Royal Society of Canada reviews of studies conclude that United States' positions cannot be reconciled with the agreed text.

June 1983: Series of studies on acid rain from U.S. groups, including the U.S. National Academy of Sciences, confirms most of the science and points advocated by the Canadian government.

March 1984: Federal and provincial environment ministers from Manitoba east agree to a 25 per cent reduction in acid rain–causing emissions by 1990 and a further reduction of 25 per cent by 1990 if the United States undertakes similar action.

March 1984: Under Canadian leadership, Canada and nine European countries sign a declaration of intent to reduce SO_2 emissions by at least 30 per cent by 1993. This '30 Per Cent Club' is later expanded in July 1985 to include twenty-two countries. The United States and Britain decline to join.

February 1985: The federal government and the seven easternmost provinces agree to reduce annual emissions of SO_2 by 2.3 million tonnes, a 50 per cent reduction from 1980 allowable levels, by 1994.

TABLE 4 (continued)

March 1985: The Mulroney–Reagan 'Shamrock Summit' acknowledges the commonality of the acid rain problem and appoints special acid rain envoys to develop cooperative solutions.

March 1985: Cabinet approves the Acid Rain Abatement Policy, including financial support for smelter modernization.

December 1985: The Ontario Peterson government announces an $85-million 'Countdown Acid Rain' program to cut SO_2 emissions in the province by 67 per cent.

January 1986: The Special Acid Rain Envoys, Drew Lewis and Bill Davis, release their report on acid rain. It recommends a $5-billion program to research more efficient control technologies. But the United States continues to oppose further emission controls.

September 1986: The U.S. Court of Appeal overturns an earlier lower court ruling that would have compelled seven states to reduce acid rain–causing emissions.

March to October 1987: Agreements are signed between Canada and the provinces of Newfoundland, Prince Edward Island, New Brunswick, Manitoba, Ontario, and Quebec to implement commitment to reduce sulphate emission levels by 50 per cent of their 1980 levels.

March 1987: Canada, Quebec, and Noranda Inc. sign an agreement which will enable Noranda to reduce its SO_2 emissions at its Rouyn-Noranda copper smelter by 50 per cent by 1990.

November 1987: The United States signs an air-quality agreement with Mexico. In this case it is Mexico that is emitting the pollutants causing damage.

February 1988: Canada and Nova Scotia sign an acid rain–control agreement.

November 1990: The new U.S. Clean Air Act is signed into law by President George Bush. It contains a U.S. program to reduce acid rain–causing emissions.

While the early pollution focus in the DOE was clearly on Great Lakes water issues, it was recognized from the outset that air pollution would have to receive more concerted attention. This attention would, in turn require more of a research focus within the AES.

The AES, as chapter 1 showed, was no newcomer to the Ottawa scene. Its core was the 120-year-old Canadian Meteorological Service. It was certainly a technical service, but it was not known as a research organization per se. In 1971, coincident with its move to the DOE, the Atmospheric Research Directorate was formed to research and model atmospheric processes, air quality, and interenvironmental relationships and physical modelling. Thus, from the outset the AES was involved both in research on acid rain and in ensuring that research being done elsewhere in the world was made known to DOE regulators. The AES also led a coordinated federal-provincial program of research and monitoring of the impacts on lakes, ecosystems, and structures as well as on atmospheric aspects.

Another source of both pressure and activity on the acid-rain front came from the Air Pollution Control Directorate. As shown in chapter 4, this directorate had quickly become the key player in operationalizing the Clean Air Act, a task it performed through a basically collaborative approach with the provinces. Among the scientific and professional levels of the DOE, the concerns about acid rain were beginning to surface as early as 1973.[3] Some of the early information through the work of the OECD, where an LRTAP committee had been formed. That work had been precipitated by Norwegian and Swedish research. Early efforts by DOE professionals to express the seriousness of the environmental consequences got nowhere. As one DOE scientist put it, 'I remember one senior Ontario Hydro official totally and publicly belittling me when I tried to raise the topic at a conference in 1976.'

But scientific impetus was growing. For example, in May 1975, a major international symposium held in Columbus, Ohio, brought the seriousness of the international research findings, including their application to North America, to broader light. The Science Council of Canada had launched its 'policies and poisons' study, which included an examination of acid rain, particularly at the insistence of its study coordinator, Dr David Bates.[4]

By August 1976, there was sufficient impetus within the DOE for the deputy minister to establish an integrated program on LRTAP. The primary purpose of this mainly research-oriented program was to determine the baseline state of the environment in Eastern Canada, especially prior to the impact of coal-burning emissions that were then projected to increase in the United States. A second purpose of the program was to understand the transport, occurrence, and effects of these pollutants, including geographical extent, severity, and socio-economic costs.

In 1977–8, two events occurred which moved the acid-rain dossier up another notch in the DOE priority list, in this case organizationally and politically. First, Dr Bob Slater came from Ontario to be the acting director general of the Air Pollution Control Directorate. Because acid rain had been more important in Ontario, Slater gave it a new focus within the DOE. He was thereafter to be continuously associated with the issue, taking charge of the dossier wherever he went as he advanced up the DOE ladder. But, in 1977, his actions were simply to increase the profile of the issue. He had a very tough speech written and then persuaded Romeo LeBlanc to give it at a June 1977 conference of the Air Pollution Control Association. LeBlanc knew little about acid rain at this stage but was prepared to advance the issue politically if he could. The speech was eventually toned down from the early drafts, but it did contain a phrase that got maximum media attention. LeBlanc described acid rain as an 'environmental time bomb.'

The key 1978 event, more important in the long-range acid-rain story, was the passage in November 1978 of a resolution by the U.S. Congress. Through it, the United States was effectively offering to sit down with Canada to discuss its concerns about Canadian air-pollution emissions that were impacting on the United States. The specific impetus for some congressional representatives was their concern that a coal-fired power plant to be constructed at Atikokan in Northwestern Ontario might spread acid rain into Minnesota. Some DOE regulators believed that the real U.S. concern was not pollution per se but rather the fact that the new plant would not accord with U.S. pollution standards. Hence, the U.S. concern was with ensuring that U.S. regulations applied to Canadian plants as well, in effect with ensuring that U.S. law was applied extra-territorially.

In any event, Canada jumped at the chance to discuss acid-rain pollution with the United States. The result was that a bilateral research consultation group, led by AES scientist Gordon McBean and an Environmental Protection Agency (EPA) specialist, was set up to integrate North American research on the issue of LRTAP. It is important to note at the outset that this special bilateral mechanism was chosen rather than the more independent and arm's-length International Joint Commission or the Great Lakes Water Quality Board. The latter was the model which Canada preferred to use. The first report of the technical bilateral group came out in October 1979 and pronounced acid rain as the problem of greatest mutual concern. It was this report which landed on the desk of Conservative environment minister John Fraser, with its vivid accompanying media-driven imagery of dying lakes.

While U.S. congressional players lost considerable interest in the issue when it was evident that acid-rain pollution was a two-way street, the United States was none the less a signatory to the November 1979 Convention on Long-Range Transboundary Air Pollution. Canada was also a signatory, along with other member countries of the U.N. Economic Commission for Europe. This convention gave a further symbolic boost to the acid-rain issue. It lacked an enforcement mechanism, but it did commit its signatories to cooperate in monitoring and research and in the joint development of control strategies.

While research continued into acid rain, the main Canadian concern in 1980 was a lingering view that acid rain might easily fall off the agenda. Thus, in July 1980, the Coalition on Acid Rain was formed. A key mover in its formation was John Fraser, by then back in opposition and concerned that the issue which he had done much to give a high political profile was going to be ignored by the restored Trudeau Liberal regime. At this stage Fraser saw the coalition as being needed more in the domestic than in the U.S. political arena, where it later gained deserved fame for political tenacity and skill.

But it was lagging U.S. interest in acid rain that was concerning the DOE the most. During 1980, Canadian strategy was to press the United States to move closer to some kind of agreement for joint action.[5] In August 1980, with the Carter Adminstration still in power, Canada secured the signing of a Memorandum of Intent which set out the mutual goal of the two countries to take concrete cooperative steps to combat acid rain. The memorandum also called for the development of domestic air-quality laws, increased notice of proposed actions, and increased research and monitoring. In December 1980, Canada amended the Clean Air Act to include international air-pollution control provisions similar to those in the U.S. Clean Air Act. Through the quick passage of this amendment, it was hoped that the proper statutory provisions were in place as a vehicle to bring the two countries together to develop a joint solution.

Despite these various bilateral and multilateral statements and understakings, the situation by June 1982, when Canada's acid rain talks with the Reagan Administration broke down, was still in the realm of research and mutual cajoling rather than actual control action. This situation also applied domestically. Canada was actively discussing with the provinces what kinds of actions Canada could take to reduce acid rain-causing emissions, but these too were at the symbolic stage.[6] In April 1981, the federal government and four provinces, but not including the key province, Ontario, announced that they would commit themselves to holding acid-deposition loadings to an initial target of no more than 20 kilograms per hectare per year by 1990. While there was pride at the DOE in the AES science role that underpinned the acid-rain dossier, the rest of the 'getting ready to get ready' symbolism was becoming both embarrassing and frustrating.

Bilateral Stalemate and Reagan-Style Hardball Politics

The period from June 1982 until the end of 1989 was marked by continuing stalemate regarding the vital Canada–United States relationship on acid rain but a breakthrough regarding domestic action within Canada. Explanations for both the failure and the success reside in the way political coalitions congealed or failed to be mobilized on both sides of the forty-ninth parallel.

Several elements of the U.S. political-economic equation made acid rain a far more difficult environmental issue to resolve than earlier major concerns such as Great Lakes water pollution. One of these was hinted at in chapter 6, where we noted how the U.S. state governments coalesced around the Great Lakes issue. In the water issues, the states that were the source of the problem were also the victims of their own pollution. In the case of acid rain, the sources of the pollution, mainly coal-fired power plants in the midwestern states, such as Ohio, and in West Virginia, formed a politically distinct coalition, whereas the victims of the pollution were in the New England states, New York, and to some extent Pennsylvania, and stood in opposition.

A second complication was that there was a difference in the primary causes of the acid rain in the two countries. For Canada it was primarily smelters, and for the United States, it was coal-fired power plants. This difference in turn affected the nature of the coalitions involved. In Canada the smelter owners were a small handful of key firms such as INCO, Ontario Hydro, Falconbridge, and Algoma. In the United States the producers were not only the more numerous power-generation companies but also the coal interests, which had considerable muscle, both company and union.[7] These differences also influenced the nature of control approaches and mutual arguments about environmental virtue. For example, regulations under the U.S. Clean Air Act required the installation of expensive scrubbers on U.S. coal-fired plants. Canada had no similar requirements, a fact which the U.S. interests used to argue that Canada was insincere in its approach. For Canada, it was smelters that were the problem, with each of the main plants having a different production mix and technology.

In any event, the coalitions at the state level were quickly replicated and reinforced at the congressional level in Washington and thus, from the outset, promised to make acid rain a tougher political nut to crack than previous Canada–United States environmental problems. None of these inherent dynamics was made any easier in the early 1980s by the fact that economic times were difficult. The 1982 recession had bludgeoned the same midwestern areas of the United States, transforming some of the same coal-fired power states into the so-called rust belt as thousands of manufacturing jobs were lost. It was not difficult for U.S. interests to get a receptive hearing in Congress to delay or oppose any control actions that might lose, or even threaten to cost, more jobs.

U.S. coal interests and power companies pulled no punches. They not only campaigned against acid-rain controls but also opposed Canadian electricity exports, and indeed referred to Canadian demands for tougher acid-rain controls as a kind of Trojan horse for such expanded electricity exports from Canada. The political heat was such that at one meeting in Cleveland, where a senior official from the DOE was scheduled to speak, a SWAT team armed guard had to be present because of a threat from one of the U.S. coal miners' unions. At another meeting in Boston, the same official was peppered with essentially hostile questions after a speech outlining the Canadian position. Diplomatic niceties were not to be a part of acid-rain politics. A frustrated DOE even hired an American journalist to write articles to spread the word about Canada's concerns. One of his successes was that indirectly he was able to deliver the public support of the U.S. National Rifle Association to Canada's side, a lobbying odd couple if ever there was one.

But all these factors paled into insignificance when confronted by the ultimate determinant of the U.S. position in the 1980s – namely, the unbridled anti-environmental attitude of the Reagan Administration, especially during its first term of office.[8] Experienced DOE and External Affairs officials could find few if any precedents for the undisguised belligerence of the U.S. position. This position was spearheaded by Reagan political operatives, but it even extended into the Environmental Protection Agency itself. Unlike the previous Carter and Nixon administrations, the Reagan team offered, in the eyes of senior Canadian players, not even an intention to cooperate after it inherited the Memorandum of Intent. The Reaganite tactics ranged from employing frequent changes in their representatives on joint committees to calling meetings that did not occur, to unabashed political censorship of scientific reports. The last of these refers to conclusions to reports made by joint technical working groups in which Reagan political operatives would produce summary sections contradicting the consensus reached by Canadian and U.S. scientists set out in the body of the reports.

The anger, frustration, and surprise at these hardball political tactics were best reflected in a speech given by Ray Robinson to the U.S. National Academy of Sciences in October 1982. Robinson had just left as head of the EPS in Environment Canada. The speech, though cleared politically, could barely contain itself within the normal rules of diplomatic niceties. Robinson concluded with the following statement:

For three and a half years I headed Canada's Environmental Protection Service ... I saw, as a non-scientist, that most questions facing government could not be resolved by science. But I also saw that scientists, if allowed to proceed to the limit of their knowledge and to draw their conclusions in an independent fashion, could often significantly narrow the areas where non-scientific judgements had to be applied. I also saw how dangerous it was to seek to influence scientific judgements to produce politically or administratively convenient conclusions. With such an approach it would not be long before it would be impossible to distinguish between fact and fancy. I fear that in parts of Washington that has already happened with respect to acid rain. [9]

The more that the acid-rain battle became a political struggle, the more it left the domain of the AES and became the primary concern of the EPS. Between 1980 and 1982, the dossier occupied as much as half of Ray Robinson's time as head of the EPS. Day-to-day management increasingly resided in the capable hands of Alex Manson, who stayed with the issue throughout the 1980s. But Canada was virtually powerless in the wake of U.S. intransigence, not because the DOE itself was weak but rather because Canada had few effective levers over the U.S. giant.

Actions in 1983 and 1984 brought out this sense of fundamental political incapacity. In February 1983, the reports of the bilateral work groups were released but again revealed the divisions regarding the aforementioned political conclusions. For tactical reasons, the DOE asked the Royal Society of Canada, Canada's pre-eminent scientific society, to review the reports of the three bilateral work groups. The Royal Society panel concluded that the Canadian proposal followed from the *agreed* text of the work-group reports but that the U.S. version could not be so reconciled with the actual report.

Early in 1984, Canada also secured the formation of the '30 per cent club.' As chapter 6 showed, Canada and nine European countries signed a declaration of intent to reduce their SO_2 emissions by at least 30 per cent by 1993.[10] In a multilateral context, the '30 Per Cent Club' process was reflective of several efforts by Canada to secure overall international environmental progress and influence. But as a specific event in the total bilateral acid-rain story it is more reflective of political weakness than anything else. The '30 Per Cent Club' was forged to get at two different acid-rain enemies, the United States, in the case of Canada, and the United Kingdom of Margaret Thatcher, in the case of the European, especially the Nordic, countries.[11] And while it made for good symbolic politics, it had little effect on actual U.S. or British behaviour regarding acid-rain controls.

Until 1984, the Canadian strategy on acid rain had been premised in part on a view that Canada would act on about half its control program in concert with, and to the extent that, the United States did likewise. There was considerable sense to this strategy not only in bargaining terms but also because, for some key areas of acid rain, such as in the Muskoka area of Ontario, 70 per cent of the fallout was coming from the Ohio Valley. Thus, even a strong Canadian control program would have only limited positive effects in some areas.

For its part, the United States adopted the strategy of continually arguing that Canada had to show scientifically that there was in fact a problem. Moreover, Canada also had to show good faith by enacting its own control strategy or Congress would not be much impressed. Despite the profile of the issue, there was not enough domestic political support in Canada to make a Canadian control program stick. In a perverse way each country was playing a game of reverse environmental chicken by inducing or forcing the other to show its environmental *bona fides* first.

The United States won this tactical struggle, and by the end of 1985 Canada had announced a federal-provincial control program in concert with the key producer companies. But an account of the nature of this domestic breakthrough requires a more detailed look at the positions adopted by the key provinces and corporate players and at some of the inherent economics of control.[12]

Domestic Breakthrough

In February 1985, the federal government and the seven easternmost provinces announced that they had agreed to reduce annual emissions of SO_2 by 2.3 million tonnes by 1994, a 50 per cent reduction from 1980 allowable levels. A month later, the federal Cabinet approved an acid-rain abatement program including a $150-million financial support package for smelter modernization. By the end of 1985, the Ontario government had launched its own $85-million 'Countdown Acid Rain' program, matched to an equivalent federal amount, which would cut emissions in the province by 67 per cent.

At first glance, the most obvious reason for seemingly sudden action was that there were two new governments in power, in Ottawa and Toronto. The Mulroney Conservatives had assumed office in the fall of 1984, and in the 1984 election campaign Mulroney had promised a comprehensive acid-rain program within six months of taking office. In early 1985, the new prime minister was preparing for the Shamrock Summit, his first summit meeting as prime minister with U.S. president Ronald Reagan. The acid-rain initiative was being orchestrated by the Prime Minister's Office, with the DOE's willing cooperation. The Mulroney agenda had stressed that the Conservatives would establish a new, close and constructive relationship with the Americans to set their government apart from what they regarded as the excessively belligerent approach taken by the Liberal government in the early 1980s.[13] The Summit would pave the way for the launching of free-trade negotiations as well as indicate further that Canada was 'open for business.' The one issue in which the prime minister hoped he would have leverage over Reagan would be acid rain. And to keep pressure on the Americans, he wanted to be able to say that Canada had launched its own control program, thus removing the only remaining moral-high-ground counter-arguments the Americans could hopefully marshall.

Also newly in power in Ontario was the Liberal government of David Peterson. Replacing a forty-year reign by the Conservatives at Queen's Park, the Liberals were anxious to show their environmental credentials, and in the process put pressure on the new federal Conservative government as well. The Countdown Acid Rain program was the province's showpiece and marked a considerable improvement over past Ontario actions.[14]

While the imperatives of summitry and of the priorities of the newly empowered are undoubtedly important, it is doubtful that such initiatives could have occurred until the ground had been prepared for bringing the key companies along into a cooperative endeavour. Both the domestic breakthrough and federal-provincial cooperation were very much linked to the politics and economics of changing corporate production approaches, especially in the smelter industry. Ontario, in particular, especially under the well-entrenched Davis Conservative government, would not push the main offending companies until the political-economic ducks were properly aligned.

Ontario was the key to the acid-rain puzzle, but its delays in action were a function of several factors. First, especially in the media hothouse of Toronto, environmental issues always involved a game of partisan one-upmanship between the governing parties in Ottawa and Queen's Park. In the early 1980s, this involved the Tories in Toronto and the Liberals in Ottawa. In the rest of the 1980s, it was the reverse. Neither appreciated being manoeuvred into a position where they looked like the main offending environmental player. Such competitiveness was sometimes purely symbolic and utterly unproductive, but sometimes it also helped advance the acid-rain agenda.

For example, in November 1980, Ontario environment minister Keith Norton convened a meeting of provincial ministers to help get the acid-rain file moving in the federal-provincial sphere. It helped generate momentum that led to the provinces announcing target cuts. But less than a year later, Norton was protesting strongly to Ottawa that he had been 'betrayed' by federal minister John Roberts's statement to the media in New York that there had been agreements with the provinces involving 50 per cent cuts in Canadian emissions.[15]

While this sparring continued throughout the acid-rain story, it also reflected an important philosophical conflict over approaches to air-pollution control between the Ontario and federal environment departments. The Ontario approach had been to utilize the air as a resource for the dispersion of pollutants.[16] It therefore sought to establish controls at the point of impingement. In short, controls were used only as needed to keep ambient air concentrations below harmful levels at the nearest point, where the emissions hit the ground or a building. This approach presupposes good scientific knowledge of harmful levels of pollutants. The construction of the INCO tall stack in Sudbury and similar U.S. actions were based on this control philosophy.

The federal approach, on the other hand, has focused on control at the point of emission. It advocates, especially for new plants, control of the emission at source through the application of the best-practicable control technology. This approach assumes that the regulator may not have complete knowledge of the effects or behaviour of pollutants. Experts in each jurisdiction recognized some of the desirable features of the other's approach, but none the less there was a conceptual divide.

One of the effects of the Ontario approach was that Ontario had been reluctant to ask or pressure key firms such as Ontario Hydro and INCO to adopt best-practicable-technology approaches, favouring instead reductions that involved minimum costs. There is no doubt that these conceptual factors influenced the pace of response to acid rain, but it seriously underplays the sheer political clout that many of the key firms had in the inner political counsels of the Ontario government. It was the Ontario government which in turn advocated a 'steady as you go' approach regarding the key firms.

Ottawa, meanwhile, was not unsympathetic to a slow, steady approach in the early 1980s. First, it wanted its control actions to proceed with, not ahead of, U.S. actions. Second, knowing that several provinces had to be involved, the DOE did not want to scare any province away from the table. This was another reason for the use of public target-setting statements and for adopting an approach in which the commitments of one province could put pressure on others less willing to move. For example, not long after the December 1980 meeting convened by Keith Norton, Quebec's environment minister committed his province to a 40 per cent reduction in emissions. In addition, if some of the targets seemed a little too Draconian or threatening, DOE officials would stress that movement towards them might in fact already be occurring through ostensibly non-environmental policy actions already under way. Thus, certain of the conservation and energy-conversion measures that were a part of the huge 1980 National Energy Program (NEP) were pointed out by the DOE as being supportive incentives. Therefore, they argued, the remaining distance to travel was not as great as it looked.

But the key was still the smelter companies and Ontario Hydro's coal-fired generating stations. Among the smelter companies, INCO, Falconbridge, and Algoma were the largest emitters. These four firms in total accounted for 71 per cent of the emissions in Ontario in 1983. In the late 1970s and early 1980s, these firms basically opposed most control initiatives or argued that they were not feasible 'at this time.' If either the federal or the provincial environment department got too pushy or aggressive, they would both oppose directly and call on their allies in other line departments to defend them.

One of the early events that helped pave the way for the control programs launched in 1985 was a conference held at Queen's University in 1983. Convened to look into the state of the metals industry after it had been ravaged by the 1982 recession, the conference unexpectedly led to a suggestion that a special cooperative study be carried out to look at the non-ferrous-metals industry's competitive status 'consistent with long standing economic development and environmental goals.'[17] The study involved participation from the industry; from several federal departments, including the DOE, EMR, DRIE; and from the provinces. While not designed in the mode of the DOE's later Niagara consultative mechanisms, the study none the less produced a less confrontational arena for discussion than had existed before, in part because environmental issues were being linked directly to new technological options, and in part because the industry was more frightened of its competitive prospects and had to modernize its production concepts.

The study confirmed that the industry was in tough shape. But if the industry knew that it had to modernize, it also made clear to Ottawa that if the federal government wanted to piggy-back environmental progress with a modernization program, Ottawa would have to offer real financial incentives. The industry study was at this state of play late in 1984 when the new Conservative government was getting ready for the Shamrock Summit. But freeing money up from a new Mulroney government committed to a strategy of expenditure control did not make it look likely that an acid-rain package would materialize. It was ultimately the clout of the PMO in the context of the Shamrock Summit imperatives that garnered the funds.

It is also at this stage that one must appreciate the different positions of each key company. Falconbridge had recently modernized its production plants, and hence its argued position, environmentally, was that it had already acted and it was up to others to catch up. From Quebec, Noranda Inc. argued that it could afford nothing and that federal assistance would be needed. Moroever, the control program would have to be custom-designed for the circumstances of the Noranda operations in Quebec. INCO was initially quite sceptical of the whole prospect of change but eventually became a quite enthusiastic participant once they saw that they could economically take advantage of some of the newer technologies involved.

As always in these situations, final solutions turn on the widely varying situations that different firms face and whether control strategies can be flexible as to how things can be achieved.[18] The trick is also one of designing the right mixtures of carrot and stick, plus steady persuasion, so as to yield a program of action that is feasible and effective. Often such packages do not conform to what political rhetoric requires, where the language of gladiatorial battle and defined heroes and villains seems too frequently to be necessary.

But even the above portrait does not complete the picture of the abatement-control program constructed by the DOE. As the plan reveals, and as subsequent announcements showed, other provinces and their key firms also had to be brought into line with varying degrees of willingness.

The Acid Rain Abatement Policy approved by the Cabinet in March 1985 was essentially a policy for Eastern Canada (but encompassing Manitoba) and included federal-provincial agreements on emission reduction and cost sharing; a $175-million program of incentives for smelting companies; $70 million in research and development funds for reducing emissions from coal-fired facilities; and $18 million annually for acid-rain research and monitoring. The plan for smelters incorporated company-by-company modernization agreements which included technology-demonstration initiatives to improve productivity and reduce the cost of pollution control; new emission-control limits and compliance schedules; and financial-assistance frameworks.[19] Ottawa accepted responsibilty for the lead role regarding the smelters, while the provinces had the lead role regarding power plants.

While the control program was thus launched, it took another three years for all the individual agreements with the provinces and the companies to be worked out and announced. In terms of volume, Ontario and Quebec were the most important, and deals were worked out by 1987. More reluctant provinces such as Nova Scotia held out until 1988 and have been slow to comply ever since. In some instances the foot-dragging reflected deliberate political tactics. But there were also some genuine disputes that had to be resolved technically and politically. For example, the atmospheric modelling worked out by the AES would readily yield a general scenario of how loadings ought to be reduced across the country. But some provinces and companies saw these as extremely unfair in either percentage or technological terms, and hence further negotiations were needed.

U.S. Action at Last

At the Shamrock Summit of March 1985, Prime Minister Mulroney and President Reagan agreed on the 'commonality' of the acid-rain problem but, in the absence of U.S. action, announced that special acid-rain envoys would be appointed to develop cooperative solutions. Ronald Reagan would still not move on control actions, a strategy he successfully carried out during his entire eight years in office. Senior U.S. official Drew Lewis and former Ontario premier Bill Davis were the envoys, and they dutifully reported in January 1986.[20] The report was roundly criticized by environmentalists, and privately by many DOE officials as well, as just another delaying tactic. In the report the United States continued to oppose any concrete control program. But the United States did commit itself to a $ 5-billion program to research more efficient and clean coal-control technologies. In the view of some DOE officials this was a key prior step in influencing many U.S. senators and congressional representatives. When many swing voters in Congress indicated in 1990 why they were supporting the new U.S. Clean Air Act and its control program, they cited the fact that the Administration had tried to develop viable technological options and had not just bludgeoned producers. The report of the special envoys also called for a special bilateral advisory and consultative group. This too was viewed by some as a necessary, though unspectacular, step to keep the two sides talking and to keep the decision process on track. In this sense, the resolution of difficult environmental problems, especially where there are discrepancies of political power, can be likened to a long, tiring marathon race. One's willpower and determination to keep moving have to be greatest when exhaustion and frustration are setting it.

Eventually, in November 1990, President George Bush signed the new U.S. Clean Air Act into law.[21] It contained at long last a significant U.S. control program. The transformation of the U.S. position was attributable to several factors. First, Bush himself, under some of the same international pressures that propelled the Canadian government to launch its Green Plan (e.g., fears about global warming, and major environmental accidents in 1989 and 1990) pronounced himself to be an environmental president. Second, the Democrats had gained control of the Senate and the new majority leader, George Mitchell, was from Maine and was extremely concerned about acid-rain damage to his own state. Third, the cumulative pressure of Canadian governments had some impact.[22] It is also important to stress that some interests in the United States were pressing for environmental reform in general and action on acid rain in particular. For example, court action in the United States was taken that sought to compel seven states to reduce acid rain–causing emissions. Alas, a favourable lower court decision was overturned in November 1986 by the U.S. Court of Appeals. Paradoxically, the United States did sign an air-quality agreement with Mexico as early as November 1987. In this case it was primarily U.S. concern about Mexican emissions coming into the U.S. southwest.

There is a sense in the whole acid-rain story that ultimately the United

States moved when it bloody well felt like it and that Canadian pressure amounted to only a marginal form of leverage. In this regard, the attribution of influence to the actions of entities such as the Coalition on Acid Rain would seem to be delusions of Canadian grandeur. While the determination of influence is a notoriously inexact art, it would be a mistake to underplay not only the coalition's role but also the extent to which the acid-rain issue received remarkably broad and continuous non-partisan support in Canada.

As noted earlier, the initial impetus for establishing the coalition came from John Fraser interacting with his ENGO colleagues and concerned that acid rain might be dropped from the domestic agenda in 1980. But the coalition eventually became best known for the way in which it tenaciously lobbied in the United States among congressional leaders, Administration officials, and environmental groups. Headed by Adele Hurley and Michael Perley, the coalition was viewed by DOE strategists as a distinct asset in their U.S. lobbying activity. But they did not view them so enthusiastically in their Canadian lobbying effort. In their activities in the U.S. capital, they were continuously widening the knowledge and understanding of Canadian concerns and building useful coalitions among U.S. interests. In this regard, the DOE regarded the money it spent on the coalition (mainly indirectly, through contract projects) as money well spent. As for the coalition's domestic activity, DOE strategists as often as not viewed the 'Hurley–Perley' brigade as a nuisance. DOE operatives felt that the coalition often undermined them in dealing with the Americans by stressing the inadequacies of the Canadian domestic control program. This was partly true, but in the final analysis what the coalition was doing to the DOE and the Canadian government, Canada was trying to do to the United States.

The coalition put pressure on successive environment ministers, federally and provincially, and lobbied the leaders and environmental critics of the opposition parties. There is little doubt that the coalition's activity, but also the DOE's own information activities and selling of the acid-rain issue, helped keep acid rain high in the opinion polls as Canadians' number-one environmental priority for most of the decade.[23] The strong view held by Canadians was also what undoubtedly helped members of Parliament to take what was primarily a non-partisan approach to the issue.

This was reflected in several usually unanimous House of Commons committee reports throughout the decade. In 1980, the Standing Committee on Fisheries and Forestry established the Sub-committee on Acid

Rain. Its 8 October 1981 report, *Still Waters*, both urged stronger domestic action and kept pressure on the U.S. Administration.[24] This subcommittee reported again in June 1984 in a document given the unambiguous title of *Time Lost: A Demand for Action on Acid Rain*.[25] In June 1985 the special Commons Committee on Acid Rain was struck, and six months later it issued a very critical review of the special envoys' report.

Last, but not least, there were other less-well-known players in the acid-rain story whom insiders repeatedly praised for their skill and dogged efforts. One was member of Parliament Stan Darling. Darling repeatedly raised questions in the House of Commons and went on every conceivable junket to Washington to lobby members of Congress. Another was George Rejhon, a DEA official in the Canadian Embassy in Washington. Through his careful preparation of arguments and analysis, and through a well-developed network of contacts, he greatly helped the Canadian case in the lobbying maze that is Washington.

Conclusions

The decisions on acid rain span a fifteen-year period. There was little regulatory or control action for a decade before the domestic breakthrough. By the time U.S. actions take real effect in the mid-1990s, twenty years will have passed, essentially to achieve a halving of the emissions and thus, it is hoped most of the adverse environmental effects. The natural concluding question is, why was the progress so agonizingly slow and begrudging? The reasons are numerous.

The level and extent of scientific and technological uncertainty for the period as a whole were considerable. The ecosystems involved were complex, and at times intractable. This was the case even though, with hindsight, there is often a tendency to view acid rain as a simple hazard compared with later issues such as climate change. The chapter has shown this scientific and ecosystem uncertainty in several ways. First, SO_2 is but one of several interacting pollutants involved in the Long-Range Transport of Airborne Pollutants program. Many hazards are in fact involved. Second, it took several years in the early 1970s for scientists to reach sufficient minimum agreement to put it on the agenda. Third, there were serious problems in simply establishing baseline data regarding the nature of the transport of the pollutant, the level of fallout, and the precise geographical dispersion. Fourth, there were differences in the sources of pollution in the United States and Canada, which in turn affected the technical control strategies preferred in each country. Fifth, there were technical differences of view as to how much of the Canadian-located pollution was coming from the United States. Sixth, there were genuine technical differences between Ontario's point-of-impingement view of regulation and the federal focus on the point of emissions. Seventh, there were uncertainties and knowledge limits regarding the AES's modelling of overall loads versus their specific provincial levels. In short, ecosystem and scientific uncertainty, though reduced over the whole period, was never fully within the grasp of environmental regulators.

Scientific issues were, however, also politicized. The Reagan Administration did most of the politicizing of science, but the DOE and provincial environmental ministries were scarcely blameless. They too were capable of playing the game in the domestic setting. Acid rain was not necessarily a partisan issue, so the politicizing of science and the sluggish regulatory response bear ample testimony as well to the power of various interests – provincial, federal, industrial, and international – to drag out negotiations through the ups and downs of business cycles and through the raising of technological obstacles. Some of this is purely and unambiguously tactical politics, but some of it reflects genuine problems and complexities and the difficulty of getting the mix of incentives and regulations right. Some of the domestic slowness was also undoubtedly attributable to inexperience and clumsiness in bringing stakeholders together and to a penchant for waiting for unambiguous public support before seriously attempting to mobilize and muscle the key interests.

Acid rain was the DOE's torch of progress in the environmental-policy process. Successive ministers held it high, did their best to advance the cause, and passed it on to the next political champion. However, for the DOE as an organization, the acid-rain issue was a mixed blessing. On the one hand, it was one of its few reasonably unifying priorities that helped build an outside constituency. On the other hand, for other areas of the department, the inherent length of the struggle, and the resources and leadership attention it consumed, distracted the DOE from an otherwise potentially more even-handed approach to the range of environmental issues, both in place and looming on the horizon.

8

The Parks Service and the
South Moresby Decision

At his first briefing after being named minister of the environment, Tom MacMillan turned to his officials and said, 'And now tell me about the Parks Service ... tell me about the fun part of the job.' The attraction of environment ministers to Parks is understandable. Over the years it looked like a more serene and pleasant part of the Department of the Environment (DOE) mission, in part because other areas brought trouble and resistance in spades, be it battling the Americans, as we saw in the acid-rain case, or fighting entrenched industries in regulatory areas, such as pulp and paper, chemicals, or mining. Parks – creating them, improving them, being seen in them – offered the chance of taking the environmental high ground more often than in other areas of the DOE mandate.

But when seen at closer range, the Parks Service, especially in the 1980s and early 1990s, proved to be somewhat less idyllic than advertised. Tom MacMillan eventually took credit for six new parks. But the park which gained him the most political credit, South Moresby in British Columbia's Queen Charlotte Islands, involved a decision that was hardly made on a bed of political roses. It showed the complex nature of the century-old Parks Service's decision process, the diverse and free-wheeling networking and pressure of local environmental lobbies and Native groups, the way parks are sources of federal-provincial financial gamesmanship and sources of constant credit claiming and blame avoidance, and the contending issues of parks protection and parks use in the day-to-day management of the DOE's largest operational and decentralized service.

To understand the Parks element of the DOE mandate, we use the two-stage approach implied in the title of the chapter. We look first at the basic evolution and characteristics of the Parks Service from 1979, when it joined the DOE in the swap for the Fisheries component, tracing it to the state of play that the service was in when Tom MacMillan took over as minister in 1986. The remainder of the account then deals primarily with how the decision on South Moresby was made and what the South Morseby case shows about the policy and resource tensions within the Parks Service and within the DOE. The chapter ends with the Mulroney era and thus does not deal with the 1993 decision by the short-lived Campbell Conservative government to move the Parks Service out of the DOE into the new Department of Heritage.

The Parks Service as Reluctant Environmental Recruit

When Tom MacMillan was in the early days of his embrace of the fun part of his job, the object of his affection was already 100 years old. Indeed, it would be at the Parks Service's main centennial event, the Canadian Assembly of National Parks and Protected Areas, held in Banff in September 1986, that MacMillan would make his first major speech as environment minister. The federal government's involvement in parks began with the establishment of Banff as Canada's first national park.[1] Parks were at various times a part of the Department of the Interior, the Superintendent of Forestry, the Department of Indian Affairs and Northern Development, and, after 1979, the Department of the Environment.

The legislative history of Parks as an integrated service essentially begins in 1930 when the National Parks Act was passed. This legislation was a key part of the political deal that established the basic resource and land rights of the Western provinces. Ottawa and the provincial governments involved negotiated a pact that gave the provinces jurisdiction over land within their boundaries, including timber, mineral, and water rights.[2] But to satisfy conservationist pressures, no mining, hydro-electric dams, or forestry development would be allowed within the boundaries of the national parks.

The history of national historic parks and sites, the heritage side of the Parks Service mandate, can also be traced to very early legislation, in this case to the Dominion Forest Preserves and Parks Act of 1911. In 1917, Fort Anne, at Annapolis Royal, became Canada's first national historic park. In 1919, the Historic Sites and Monuments Board was established to advise the minister on what would now be called heritage issues.

Thus, by the time the Parks Service joined the DOE, it had had a long history as a proud service. Moreover, in whatever department it had been located, it had enjoyed a high degree of independence. In part, it was accorded such independence simply because it is a large operational entity functioning in a decentralized way at numerous geographic locations across the country. Moreover, for most of the early part of the 1970s it had had, in its location in the Department of Indian Affairs and Northern Development, a minister who had also seen it as the most enjoyable part of his job. Thus Jean Chrétien, before he moved to more senior ministerial status, had presided over the establishment of several new parks. Indeed, in reality, it was Chrétien who invented and honed the fine political art of parks creation.

For all of these reasons and traditions, the Parks Service was an exceedingly reluctant recruit to the DOE. As Al Davidson, the then assistant deputy minister Parks, confided to one colleague, 'The Parks Service is a beautiful place to manage.' By this he meant that basically it gave satisfaction to all those involved, was a service wanted by Canadians, and was subject to minimum ministerial interference. And where interference occurred, certainly Chrétien style, it was friendly and basically encouraging. Moreover, in 1978, the service had reformed its own policy frameworks to ensure that every park was governed by a five-year management plan and that park decisions included a built-in public-participation process. The latter had helped overcome some of the criticism levelled at the Parks operation in the early 1970s that it had been deficient in integrating the parks with their local economies and cultures.

The Parks Service uniformly opposed joining the DOE because it also knew that it was entering a hornets' nest. As chapter 1 showed, senior Parks officials feared that they would become the target for the rest of the DOE, anxious to recoup budgetary and personnel losses accompanying the departure of the Fisheries component. The Parks Service would constitute half of the DOE's budget and personnel establishment, but its capital budget was an especially inviting target since raids on it seemed to have the virtue of not immediately affecting current operations. These fears were quickly realized. The DOE's 1979 document on priority issues for the 1980s noted that one of its priorities would have to be the 'restoration of the Parks Canada capital budget.'[3] The document noted that the Parks capital budget had been cut from $140 million to $82 million. But, alas, this priority for the DOE was also shown unambiguously as being near the end of a fairly long list. In 1982–3, an extensive review of the Parks budget, contrary to most such reviews in Ottawa, showed that the service was 'under-resourced by about 300 person-years to maintain and operate the existing parks and sites.'[4]

Protection versus Use, and Old Parks versus New

The mandate of the Parks Service is 'to protect for all time those places which are significant examples of Canada's natural and cultural heritage and also to encourage public understanding, appreciation, and enjoyment of this heritage in ways which leave it unimpaired for future generations.'[5] Thus, contained within the Parks mandate is a built-in real or potential conflict between the goals of preservation and user enjoyment. The national parks system at the time of the early stages of the South Moresby decision consisted of 31 parks covering 140,000 square kilometres, a bare 1.3 per cent of Canada's land mass.

Parks Canada had established, following public consultation, a larger plan for Canada's eventual parks system. It was based on the achievement of representative parks for each of the thirty-nine natural terrestrial regions. In the mid-1980s, there were parks in twenty of these regions. Eleven regions in the South and eight in the North still required protection. In addition, marine parks were also envisioned. The DOE Strategic Plan for 1984 to 1989 showed that the department was committed to the 'establishment of ten new National Parks by 1990.'[6] The main criteria for the establishment of new parks were first, whether they fit within the representative plan and, second, whether a given area was subject to an imminent threat that would prevent the preservation goal from being realized.

But given the ministerial penchant for creating new parks, and given increasing budgetary restraint in the 1980s as a whole, there increasingly developed within the Parks Service a conflict between new parks and old parks. The auditor general's review of Parks criticized it for being ad hoc and opportunistic in the formation of new parks and chastised it for being neglectful of the needs of existing parks.[7] The Parks Service candidly pleaded guilty on both counts.

The leadership of the Parks component knew that a delicate juggling act was inevitably a part of the territory. Ministerial covetousness for being identified with, and remembered for, a new park did advance the cause of completing the larger representative Parks plan. At the same time, the larger plan disciplined ministers, stating that not just any area could be designated a park and that not all of the area within a park had to be preservation-oriented. At the same time, every time a new park was established, its start-up costs and longer-term resource needs would automatically mean fewer resources for existing parks, whose capital and operating budgets were also being coveted by other parts of the DOE.

As if the conflict between protection and users and between old and new parks was not enough, there was also tension within the Parks Service between the natural and the heritage aspects of the mandate.[8] The historic-sites area of the service often felt itself to be vulnerable and ignored within the Parks operation in much the same way that the service as a whole perceived itself to be at risk within the larger DOE enterprise. When Jim Collinson became the assistant deputy minister for the Parks Service in 1986, he made a special effort to integrate the heritage and historical and cultural aspects of the mandate with the larger Parks program. Indeed, in 1986 be was elected chairman of the UNESCO World Heritage Committee.

A further glimpse into the kinds of mandate and resource pressures that were coalescing around the Parks Service as the South Morseby saga began to unfold can be found in the nature of the recommendations made by the Nielsen Task Force review team in 1985. The report stressed that national parks were 'exceptionally strongly supported by the Canadian public'[9] but that there were ever-increasing pressures developing around the parks system. In particular, it cited continuous pressure by the federal agency Tourism Canada; the private-sector tourism lobby; and most of the provinces, which argued that Parks Canada was 'misinterpreting its mandate by an overemphasis on protection'[10] and by an underutilization of the tourism and economic potential of the parks. Native Peoples were also increasingly critical as parks increasingly came head to head with Native land claims and defence of traditional ways of living and earning an income from the land. This was especially the case for the new parks planned for the northern parts of Canada.[11]

While conceding that the primacy given to protection was popular among Canadians as a whole, the Nielsen team none the less argued that the way ahead for the Parks Service pointed in the direction of cooperative management and financing of the parks. It advocated an approach which would shift the thrust of national parks to a position of shared ownership and joint management with the provinces, with non-profit groups, and with Native Canadians.[12] For many, this line of argument was unambiguously a pro-tourism posture.

The views that brought out the main double axes of tension in the mandate – protection versus users, and old versus new parks – were reflected in various actions and statements in the mid-1980s. For example, in an uncompromising 1984 speech to the Banff Chamber of Commerce, environment minister Charles Caccia came down hard on the side of the protectionist aspect of the Parks Canada mandate. DOE insiders regarded Caccia's views as being almost religious in their vigour. The Parks Service, in Caccia's view, was not to balance users and protection. Rather, protection was to be the pre-eminent goal.

Suzanne Blais-Grenier, on the other hand, was clearly imbued with the joint-management and joint-financing ethos advocated by the Neilsen Task Force. As we see below, in the South Moresby decision, her main offer to the environmentalists involved in South Moresby was a deal involving a three-way financial split between Ottawa, the Province of British Columbia, and an environmental public-fund raising component. She also became embroiled in a media controversy over whether she had in fact advocated the right to carry out logging and mining operations in the parks. This episode was one of the contributing factors to her ultimate political downfall in the summer of 1985.

When Tom MacMillan took over the reins of the DOE, he became quickly aware not only of the policy tensions but also of the basic instincts and stances of the main players involved in Parks decisions. Senior Parks bureaucrats were motivated by a desire to see the larger Parks plan carried out but without too many new resources going into new parks at the expense of old parks and their urgent continuing capital and operational needs. Parks management in the field, centred around the roles of the park superintendents, wanted the five-year Parks management plans to be carried out and fully resourced but with full flexibility given to their own feel for local political-economic circumstances. After all, the superintendent's job was a delicate one. It involved responsibility for a huge chunk of real estate; hundreds of employees; millions of visitors and numerous adjacent local governments, community groups, and tourist interests, each of whom saw the park as being partly theirs. And given that there was a park located in many of the constituencies of key members of Parliament, the superintendent, much like the minister, always knew that political sensitivities were never far removed from day-to-day management. But at headquarters there was an equally strong view that the five-year management plans should never be detailed and fully costed but rather should be directional and fairly general in nature.

To this internal set of players one quickly added two outside interests of no small import. One was the provinces. They had parks of their own, some adjacent to the national parks, but, for the most part, the provincial instinct was to downplay the preservationist ethic somewhat and to give greater emphasis to the tourism and user-enjoyment ethos.[13] The provinces were not themselves, however, a bundle of consistent positions. They were anxious, on the one hand, to prevent parks from being used as a vehicle for federal intervention in provincial resource-development prerogatives. In this regard, they were readily cheered on by provincial resource industries in general and by the key companies that might have timber or mineral rights in an area being considered as a park site. But, on the other hand, they knew in their political soul of souls that establishing parks was usually popular, greatly enhanced the inherent value of adjacent economic activity, and, as an added bonus, was a good way to get money out of Ottawa.

And last, but hardly least, there were the Parks ENGOs. By their very nature these tended to operate from a grass-roots base. As the South Moresby case will show, the environmental constituency on parks has some links to larger environmental coalitions, but, at its core, it is a locally or provincially based set of interests and often key individuals. The local constituencies almost invariably consist of some ardent preservationist groups, others who are anxious for the local jobs and enjoyment that ensue from a park's presence, and a middle group that straddles and mediates between the other two. Indeed, as we saw in chapter 5, on the ENGOs, the need to force these diverse groups to confront one another was among the reasons why Parks Canada introduced a public-participation process into the establishment of parks and into the development of a five-year management plan for each park. The management plan must be approved by the minister. It contains a statement of objectives, and the means and strategies for developing them.

There are, of course, some larger national ENGO organizations which are part of the Parks Service network. These include groups such as the Canadian Nature Federation, the Canadian Wildlife Federation, and the Canadian Parks and Wilderness Society, which are usually considered to be among the more conservative or moderate ENGOs. The DOE does not fund any of these groups to a significant extent, in part because, as older established groups, they usually had a larger stable membership than other kinds of ENGOs. But this decision was also part of a deliberate strategy devised by Al Davidson, as ADM Parks, not to compromise such groups with the stigma of federal funding.

Getting South Moresby on the Agenda: From Local to National Politics

The story of the establishment of the South Morseby National Park is best told, with the aid of table 5, in two phases. The first phase begins in the early 1970s and ends with the election of the Mulroney government in 1984. During this long period, South Moresby emerges from being a local and regional issue to one of national symbolic importance. The second and shorter phase, from early 1985 until July 1987, focuses more on the nature of the British Columbia–federal government negotiations and on the divisions over these negotiations within the DOE.[14] It is at this point that the Parks Service's dual policy and resource tensions, between protection and users, and between old and new parks, takes concrete form. The battle shifts from one of pure environmental politics to one that is equally over resources and budgets inside the structures of executive and bureaucratic power.

The story of South Moresby Park, or Gwaii Haanas, as the Haida Nation calls it, begins and ends with the land and the waters themselves. The glorious forests and wildlife of South Morseby on the Queen Charlotte Islands of British Columbia were central for thousands of years to both the identity and the way of life of the Haida Nation. For many of the people who became involved in the South Morseby decision and who had never seen Haida Gwaii, the Haida name for the islands, their first visit seemed inevitably to transform them. Whether it was the artist Robert Bateman, scientist David Suzuki, or Tom MacMillan and Austin Pelton, the federal and provincial environment ministers, each immediately became preservationist. They could not, of course, feel as deeply as those among the Haida Nation for whom the area was both home and history. But there is no doubt that these visitors and many other Canadians were moved by their experience on first seeing the South Moresby area.

While, in Haida terms, it is presumptuous to say that the South Morseby story begins in 1974, for our purposes it must. It begins with a growing protest led by local individuals such as Guujaaw, Miles Richardson, and the Haida elders to the prospect of increased logging in the area.[15] Rayonier Canada (B.C.) Ltd had been logging on Talunkwan Island, adjacent to the northern portion of South Morseby. In the fall of 1974, the company was holding meetings promising jobs and prosperity for Burnaby Island, farther south. A local Islands Protection Committee and the Haida Council voiced opposition to logging.

TABLE 5
Key Events in the South Moresby Decision

Fall 1974: The Rayonier logging company holds meetings in Skidegate promising employment and economic opportunity for Burnaby Island. The Islands Protection Committee and the Haida Council voice opposition to logging.

December 1974: The British Columbia NDP government announces a five-year moratorium on logging on Burnaby Island.

April 1975: The BC government grants Rayonier a licence to log Lyell Island.

Fall 1977: Federal Fisheries Department orders that logging not be allowed in Riley Creek area of Graham Island. Province and industry denounce federal interference in provincial domain.

February 1977: Local individuals take BC minister to court in effort to force a hearing on the renewal of the Rayonier logging licence being granted by the Social Credit provincial government. Court action fails.

1978: Parks Canada study leads to nomination of Ninstints for World Heritage Site designation and indicates potential for entire South Moresby area as a national park. Local groups begin to favour park idea, but Haida opposed because of failure to resolve land claims.

June 1979: Provincial planning team on South Moresby is established, involving all the key interests, but terms of reference preclude option of preservation of entire areas as wilderness area.

1980: Jim Fulton, NDP member of Parliament, introduces Private Member's Bill to establish South Moresby as national park. Joe Clark, Tom MacMillan, and John Fraser are among supporters of initiative.

1982: First of several David Suzuki CBC television documentaries on South Moresby aired.

February 1984: Environmentalists lobby Charles Caccia and gain his support.

March 1984: Charles Caccia proposes a maximum preservation option for South Moresby through a mixed provincial and national park approach. Federal proposal is later made public.

June 1984: Conservative party under Brian Mulroney indicates that, if elected, it will support establishment of South Moresby National Park.

January 1985: National Committee formed to lobby for South Moresby. It includes personalities such as Robert Bateman, David Suzuki, Jim Fulton, and Charles Caccia, and obtains resource support from the Canadian Nature Federation and the National and Provincial Parks Association of Canada.

TABLE 5 (continued)

1985: Federal Environment Minister Suzanne Blais-Grenier proposes to environmentalist that the federal government would support a park proposal costing $6 million, funded equally by the federal government, the province, and the environmental public.

March 1985: BC Environment Minister Austin Pelton indicates his support for park status in at least some part of South Moresby.

Summer 1985: Canadian Environmental Network meeting expresses strong interregional support for South Moresby. John Fraser gives private undertaking that, as Fisheries Minister, he will do all that he can to prevent logging in South Moresby.

Summer 1985: The pro-logging Moresby Island Concerned Citizens group is formed and lobbies province to protect jobs.

August 1985: Blais-Grenier is replaced as Minister of Environment by Tom MacMillan.

September 1985: John Fraser resigns as Fisheries Minister over tainted-tuna affair.

September 1985: Haida Nation announce that they will block logging if logging permits are renewed.

September 1985: The Canadian Assembly on National Parks and Protected Areas, held at Banff to commemorate the Parks Service's 100th year, becomes a forum for the South Moresby lobby. In his first major speech as environment minister, Tom MacMillan expresses his unreserved support for South Moresby.

October 1985: BC Environment Minister Austin Pelton announces formation of Wilderness Advisory Committee. New logging permits for the south side of Lyell Island also announced the same day. Haida Nation claim a betrayal of their people.

October-November 1985: Haida blockage occurs, with national television coverage. Seventy-two Haida arrested by end of November.

June 1986: Federal and BC officials meet for first negotiation sessions on South Moresby, but meeting goes nowhere. BC asks for more time to develop their position and insists that Ottawa pay off debt of Pacific Rim National Park.

July 1986: Environmentalist Elizabeth May is appointed as an assistant in Tom MacMillan's office, designated by MacMillan as his ambassador to the environmental movement and its ambassador to him.

July 1986: Several Haida renounce their Canadian citizenship as protest against lack of protection of their land.

October 1986: New BC Vander Zalm government's Cabinet shuffle results in replacement of Austin Pelton with Stephen Rogers, a former BC energy minister.

TABLE 5 (continued)

October-November 1986: Early federal-provincial negotiating sessions show divisions about how much area in park will be allowed for logging. Includes Windy Bay. Splits occur within the DOE between more conciliatory position of Parks Service and hard maximum-protection position of Elizabeth May.

March 1987: Mulroney–Vander Zalm meeting re-energizes South Moresby negotiations but still does not settle how much of area should be protected.

April 1987: BC surprises federal minister by changing bargaining tactics. BC proposes new extensive protection areas but with proviso that there would be ten more years of logging and that Ottawa would pick up most of the costs set by BC at over $100 million, almost three times the amount previously estimated by Ottawa. Ottawa was given only a few weeks to respond.

April 1987: South Moresby issue in Ottawa is taken up by Deputy Prime Minister Don Mazankowski, out of concern that dispute may harm larger federal–BC relations in constitutional discussions.

May 1987: Parliamentary motion in support of South Moresby receives all-party unanimous support.

July 1987: Federal–BC agreement signed establishing Gwaii Haanas/South Moresby National Part, including Windy Bay.

The first outside recourse for this local opposition was to appeal to the new NDP government headed by Premier Dave Barrett. The government responded favourably by announcing in December 1974 that there would be a five-year moratorium on logging on Burnaby Island. Less than six months later, however, Rayonier had switched tactics and succeeded in obtaining a licence to log Lyell Island, just north of Burnaby Island.

The beginning of a federal-provincial battle over the issue began in an essentially unrelated conflict over an issue in the fall of 1977 in the Riley Creek area of Graham Island. Here the federal Department of Fisheries had used the Fisheries Act to order an end to logging in the area because it was harming the local fishery. This intervention by Ottawa was immediately attacked by both the provincial government and the forestry industry as an act of intolerable interference in provincial affairs and resource management.

Meanwhile, in the South Moresby area itself, local individuals had begun to involve the courts in the Rayonier dispute. They took the province to court in order to force the holding of a hearing on the renewal of the Rayonier licence by the Bennett Social Credit government. This action failed but marked a further indication of the escalating tactics of environmental politics.

In 1978 attention shifted somewhat to the federal arena when Parks Canada released a study which led eventually to the nomination of Ninstints, an area within the proposed park, for designation as a World Heritage cultural site. The same Parks Canada study had also indicated the strong potential of the entire South Morseby area as a future national park. When this news filtered to the local people on South Morseby, there was almost immediate support for the idea but only in a guarded sense since it was unclear which areas would be protected and which would not. For the Haida Nation, there was little initial enthusiasm for the park idea, and for good reason. They knew that Native land-claims issues were central and believed strongly that they owned the land. They had little reason to trust white decision makers. Moreover, they knew that a park usually meant many more visitors and users.

The ethos of the Bennett government in British Columbia was decidely development-oriented, and it had extremely close relations with the forestry industry. None the less, seeing the local pressure and the growing discussion of a possible park, the Bennett government approached the issue carefully. In June 1979 a provincial planning team on South Morseby was formed, involving most of the key local interests. But its basic terms of reference precluded the option of preserving the entire area. By definition, the approach implied a significant logging component.

While the local struggle continued in various ways, the focus of the South Moresby saga moved back to a national stage during the early 1980s, when local leaders inceasingly came to the realization that allies would have to be found to serve as a counterweight to the forestry lobby. But assembling a national constituency for a regional park was a hit-and-miss affair and there was certainly no guarantee that success was even remotely imminent.

One new source of pressure came in the form of the considerable political energy of Jim Fulton, a BC member of parliament elected in 1979. In 1980, Fulton introduced a private member's bill calling for a national park at South Moresby. Such private members' bills rarely go anywhere in a government-dominated House of Commons, but among those who supported his bill were former prime minister Joe Clark; the recent Conservative environment minister John Fraser; and the Conser-

vative Opposition Environment critic, Tom MacMillan. This was not a
bad base to build on, and two of them, Fraser and MacMillan, proved to
be pivotal at later stages of the decision process. Fulton was also the key
in mobilizing the equally considerable energies of David Suzuki.
Suzuki's CBC television program *The Nature of Things* eventually aired
three shows on South Morseby. Suzuki thus added a communications
punch to the South Moresby story that helped make it a national rather
than just a local issue.

The pan-Canadian networking and lobbying that were coming to
characterize the South Moresby issue were nowhere better exemplified
than in the role played by Colleen McCrory.[16] An ardent conservationist
whose efforts had helped secure the creation of the Valhalla Mountain
Range as a provincial park in British Columbia, McCrory launched a
personal lobbying effort in early 1984 that, as much as any other factor,
secured the support of Charles Caccia. Buoyed by Caccia's interest,
Parks Canada officials also began to increase the pressure from within.
As we saw above, Caccia was already philosophically predisposed to
support a strong preservationist stance.

In March 1984, Caccia wrote to the BC government, proposing a maximum-preservation option but for a mixed federal and provincial park.
This proposal was later made public by Caccia. When the Liberal leadership race and the subsequent 1984 election intervened, the South
Moresby campaigners immediately turned their lobbying guns on the
Mulroney Conservatives. In June 1984, they secured a public commitment from Opposition leader Brian Mulroney that his party would support the establishment of a national park at South Moresby. Being wisely
sceptical of mere election promises, the lobby sought to consolidate its
pressure on the Mulroney Conservatives once they had their landslide
victory in the 1984 election. Thus, in January 1985, a national South
Moresby Committee was formed. It contained not only the high-profile
entourage assembled during the early 1980s, such as Robert Bateman,
David Suzuki, Jim Fulton, and Charles Caccia, but also some of the traditional naturalist lobby organizations, such as the Canadian Nature
Federation and the National and Provincial Parks Association, who contributed some of their organizational resources.

From Political Theatre to Environmental Bargaining

The final, shorter stage of the South Moresby decision continued with
some of the necessary aspects of political theatre, but increasingly it

involved concrete bargaining over resources, money, time, and the inevitably distracted attention span of governments and departments with busy and not easily predictable agendas, and with changing players at the helm. Increasingly the arena of decision making shifted from free-swinging interest-group coalitions to key personalities exercising power inside the political and bureaucractic executive. It was unclear not only exactly what the mix of activities within a park would be, who would pay for it, and who its real beneficiaries and victims would be, but also whether there would be a park at all.

The next phase of the South Moresby story ran headlong into the tenure of Suzanne Blais-Grenier as federal environment minister. It will be recalled from chapters 2 and 3 that the first Mulroney environment minister was anxious to show her fiscal-restraint colours to demonstrate loyalty to the new Conservative message. She told her officials to get along with the provinces and to get along with a lot less money. Early in 1985 she shocked a group of South Moresby environmentalists by offering what would be simultaneously a radical and a stingy plan for a new park. She told the group that she was prepared to see a $6-million park established. All previous ballpark estimates had indicated costs in the range of $20 million to $30 million. The problem in her plan was not only the small total amount but also that the federal government would put up only a third of the total. One-third would come from the province, and the final third would have to come from public fund raising. It is not hard to see some of the Nielsen ethos in this plan, but the environmental groups rejected the proposal as utterly unsatisfactory.

Meanwhile, at the provincial level, her BC counterpart, Austin Pelton, was expressing more hopeful signs. Having visited South Moresby, he was an increasingly sympathetic voice in the BC government. In March 1985, he indicated his support for a park in at least some part of the area. During the summer of 1985, however, the local pro-logging lobby formed an Island Concerned Citizens Group to lobby the province to protect logging jobs.

At the federal level, another occasion for the South Moresby Park lobby to shore up its position occurred at the annual meeting of the Canadian Environmental Network, which was partially funded by the DOE. The network's annual meetings became an important rallying point for the diverse ENGO groups that belonged to it. The meeting supplied a series of votes of support for South Moresby from the various regional caucuses, thus showing Ottawa its growing pan-Canadian support. During this same period, the South Morseby coalition received private assurances from John Fraser, the new Mulroney minister of fisheries and oceans, that he would use whatever powers under the Fisheries Act he could to ensure that logging would not occur. Fraser's assurances were given in part out of his growing concern at how much damage Blais-Grenier's tenure was doing to the cause of his own party's environmental reputation. At later stages in the decision, even while Speaker of the House of Commons, Fraser intervened firmly but discreetly both with the Conservative BC caucus in Ottawa and with ministers to keep the South Moresby ball rolling when it seemed likely to stop.[17]

Alas, within a few weeks, the South Moresby lobby received what for them was both good and bad news. First, it was cheered, as were all environmentalists, by the resignation of its arch nemesis, Suzanne Blais-Grenier. Then, it lost its leading warrior within the Mulroney Cabinet, when none other than John Fraser was forced to resign over a food-product health scandal involving the selling of tainted tuna.

It was difficult for everyone involved in environmental policy to estimate just what to make of the new Mulroney minister, Tom MacMillan. As we profiled in chapter 2, MacMillan was very junior in ministerial rank. His brother Charles, a key adviser in the Prime Minister's Office, was better known in Ottawa. Moreover, as chapter 2 showed, MacMillan knew that the department he headed would not get any new money. Accordingly, all his eventual priorities were in part premised on doing things that did not cost money. He also knew that, following the Blais-Grenier débâcle, he would have to do something dramatic to establish his credentials with the ENGOs.

An opportunity for just such a dramatic gesture was presented to him on a platter by two events that occurred within a few weeks of his appointment. First, the Haida Nation made known its intention to block any logging on South Morseby if logging permits were renewed by the provincial government. The logging permits were expected to be so renewed. Second, his first major speech as minister was scheduled to be given in Banff at a special meeting to commemorate the centenary of Parks Canada. The Canadian Assembly on National Parks and Protected Areas, as the event was called, also became the next rallying point for the South Morseby lobby.

The lobby and the minister needed each other, and MacMillan gave a dramatic speech, throwing his support behind the park proposal. Provincial developments again seemed to be moving in broad sympathy with federal moves. In October 1985, Austin Pelton announced the formation of a Wilderness Advisory Committee for the area to consult on overall park development. Parks Canada made a major brief to that committee. But, to Pelton's acute embarrassment, new logging permits for the south side of Lyell Island were made public the very same day as his own announcement.

The Haida Nation angrily denounced the BC logging permits as a betrayal of undertakings made to them and immediately began a blockade. The blockade received conspicuous national television coverage during October and November 1986 and, by the end of November, seventy-two Haida had been arrested. Later, in July 1987, several Haida members publicly renounced their Canadian citizenship as a protest against a country that would not protect their land.

For most of 1986 the South Moresby decision process was characterized by federal-provincial sparring and also by some further changes in the cast of players. In June 1986, at what was supposed to be the first formal negotiation session between federal and BC officials, the province asked for more time to prepare its position and also insisted that before a deal could be struck Ottawa would have to pay off the debt from the previously established Pacific Rim National Park. Arrangements on the amount of this debt had, however, only recently been agreed to. In October 1986, the new Vander Zalm government had a Cabinet shuffle which resulted in the removal of Austin Pelton and his replacement by a minister, Stephen Rogers, whose pro-logging instincts were strong.

Earlier in July 1986, another South Moresby player joined the fray, in the person of Elizabeth May. An environmentalist, May had been approached by Tom MacMillan to become a special assistant in his Ottawa office, to be what he described to her as his ambassador to the environmental movement and its ambassador to him. May accepted the offer and thereafter became a key internal player in bringing the South Moresby project to a conclusion, especially in ways that ensured the Parks package would have a maximum-preservation component. The acid test for whether a maximum-preservation goal had been achieved was, of course, the end of logging. But another telling feature was the inclusion of the key area of Windy Bay within the park boundaries.

  When further federal-provincial negotiations occurred in October and November 1986, divisions in negotiating tactics and positions became evident both between the DOE and its provincial counterparts and within the DOE. Not surprisingly, the federal-provincial differences still turned mainly on logging, with Ottawa saying that none should be allowed and the province wanting some room for it to occur. Within the DOE, the differences were mainly between the head of the Parks Service,

Jim Collinson, and Elizabeth May, with Tom MacMillan increasingly siding with May and the environmentalists she represented.

The difference in their positions fundamentally arose from how they ranked the twin tensions that characterized the Parks Service, that of protection versus user issues and that of old versus new parks. For May, there was only one issue, and that was to maximize protection. She had no sympathy for the old parks–new parks issue. Collinson, on the other hand, carried with him a different approach, part of which flowed from ingrained Parks philosophy and part of which reflected his managerial background and the task of looking after the whole Parks Service, which was inherent in his job.

Collinson had come to the Parks Service in October 1985 as ADM from a Treasury Board and managerial background, but he also had prior career experience in provincial resource management and federal regional-policy issues. He was there in part to ensure that the penchant for new parks was not carried out to such an extent that existing parks became even more severely run down than they already were. He was also guided by Parks officials, who had always argued that, for representative terrain to be present in parks, an entire area need not always be included. This concept melded nicely with the need in Collinson's mind to saw off the old parks–new parks resource split. The link here was that the larger the park and the more it emcompassed areas in which logging rights were already granted, the more expensive it would be for Parks Canada. In short, the less funding there would be for existing parks.

More precisely, Parks officials felt that, in the case of South Moresby, the whole area was not essential to represent the natural region or to make a viable national park. The natural region in which South Moresby is located was already partially represented by Pacific Rim National Park. The small, untouched watershed of Windy Bay was surrounded by lands already clear-cut on Lyell Island, and thus hardly seen as ideal by Parks professionals. It was also obvious that the cost of the maximum-protection option was going to be very high, and Parks managers did not really believe that the federal government would find over $100 million for a park. Accordingly, they were afraid that there would be no deal with British Columbia if the minister held out for the maximum area. They were also concerned that by paying very large sums for lands not essential to meet the representation goal, they would compromise the credibility and achievability of the national parks system plan in other locations.

As negotiations proceeded and became bogged down over these

entwined disputes, the familiar federal-provincial dynamic of credit claiming and blame avoidance also took hold and inevitably became caught up in the larger federal-provincial agenda well beyond the environmental-policy sphere. The South Moresby ENGOs and MacMillan were both despairing of further progress until it was learned that Prime Minister Mulroney and Premier Vander Zalm were planning a general federal-provincial meeting between themselves. MacMillan persuaded Mulroney to put South Moresby on the agenda and in short order the negotiations became re-energized.

But at the next meeting of the two federal and provincial ministers to follow up on the Mulroney–Vander Zalm initiative, Tom MacMillan was stunned by a proposal from British Columbia that called for extensive protection areas, the phasing out of logging in ten years, and a commitment by Ottawa that it would pay over $200 million to cover the tab. The BC proposal was attempting to call Ottawa's bluff because it called for a response within a few weeks. If Ottawa refused, it would get the blame for any South Moresby failure. If Ottawa accepted the deal, British Columbia would get a financial windfall, not an unimportant consideration for a Vander Zalm government anxious to reduce its deficit. Behind the scenes, some BC officials were opposed to a federal park. Some foresters and others wanted a smaller provincial park. Their hope was that the high cost of the new BC offer would scare off the federal government and allow the province to move in and save the day.

Vander Zalm was also being wooed by Prime Minister Mulroney in the context of the national leader's larger Meech Lake constitutional strategy. Thus the new BC demands were likely to get Ottawa's attention. Tom MacMillan and the ENGOs, however, viewed the new development with mixed emotions. On the one hand, the BC surprise offer opened up the possibility of greatly expanding the preservation scope of the park. On the other hand, where would they get the money from? It was a sum that was now almost three times any previous estimate. The larger amount of money needed only served to further open up the dispute between Collinson, May, and the ENGOs on the old parks–new parks issue.

The only answer in such situations for the ENGOs, Tom MacMillan, and the Parks Service was to find political godfathers in high places with access to new money. The final pieces to the South Moresby puzzle came in the form of strategic interventions, first, by Deputy Prime Minister Don Mazankowski, increasingly Mulroney's minister of everything; then by John Fraser; and later by the Dean of Tories, Dalton Camp, Mulroney's senior adviser in the Prime Minister's Office. The needed funds were found mainly in the Western Diversification program, but they were not freed up in the end, ostensibly for environmental reasons but primarily to buy needed constitutional peace with Vander Zalm.

None the less, the South Moresby lobby declared total victory. Elizabeth May called her 1990 book *Paradise Won*, and in many senses it was. The new park would include Windy Bay, and logging would be banned. On 11 July 1987, an agreement was signed between Canada and British Columbia to create the 140,000-hectare Gwaii Haanas/South Moresby National Park Reserve and a National Marine Park Reserve-both of which, however, require legislative enactment. In the agreement, Ottawa pledged to pay up to $106 million over eight years; create a regional Parks office in British Columbia; create a small boat wharf at Windy Bay; and accept that, as a matter of principle, the province would not have to bear the entire cost of any increased burden on the BC Ferry Service. Two years later an agreement was nearing completion between the federal government and the Haida Nation. The Haida would be full partners in the management of the park, through a joint-management board.

Planning and Pouncing

It is not difficult to see from the South Moresby case-study how the Parks Service can easily be caught between a rock and a hard place. On the one hand, it wants to put in place a Parks plan in an orderly fashion. But, on the other hand, it must pounce like a lion to establish a new park whenever the moment of political opportunity presents itself. Accordingly, the Parks Service also has had a strong continuing sense of its own unfinished business. In both the early and later 1980s, it launched quite long-term reviews of its policies and plans.

A strategic review was under way in 1983–4 to examine the long-term issues for 1986 to 2001. This review included an external study by an independent panel linked to a report by Professor Gordon Nelson of the University of Waterloo.[18] It identified numerous issues, ranging from the slow progress in completing the current representative terrestrial plan to issues which ultimately involved the role of the Parks Service in broader environmental policy. The Nelson report drew particular attention to the need to utilize the parks and protected areas as environmental benchmarks and monitoring sites, to enhance the gene-pool functions of the parks, and to strengthen their research role.

Organizationally, these were merely a different way of posing the question about whether the Parks Service could be or wanted to be integrated into the larger DOE mandate. The leadership of the Parks Service frequently viewed the integration issue as a no-win, one-way-street situation for them. The management of the DOE wanted the Parks resources and person-years integrated into the department but not much else. In terms of integrated policy, Parks management found the Corporate Policy Branch for the most part to be distinctly uninterested. And, as for ministerial interest in integrated policy, this was a mixed blessing and very much a hit-and-miss proposition for the reasons already set out in the South Moresby story.

Another review was launched in April 1986 but was externally generated. Concern that the Parks program was emerging from its centennial without a plan for responding to the emerging issues caused the Canadian Parks and Wilderness Society to propose that a task force be established. Tom MacMillan established a joint public and private group. Its *Parks 2000* report, published in June 1987,[19] stressed that 'the pace of progress has faltered with only three new parks announced in the past fourteen years.'[20] It noted that only three of fourteen natural areas on the Canadian Shield were represented in national parks. And the report drew attention to the fact that Canada has set aside less than 6 per cent of the countryside, compared with many other countries that are at 10 per cent. In contrast to the budgetary miracle portrayed above in the South Moresby case, the task force drew attention to the fact that in 1986 only '$2 million of the $300 million annual Parks budget is allocated for new parks and there is no long term strategy for completing the parks system.'[21]

The Parks Service, however, certainly did not stand still in the 1980s. In 1986 it introduced in Parliament a major overhaul of the National Parks Act. It included provisions for legislating wilderness areas within parks, increased authority for the protection of soil, waters, minerals, fossils, and air quality and proposed stiffer penalties for poachers. Before the South Moresby case was resolved, the Parks Service had secured the establishment of the Ellesmere Island National Park Reserve. Progress was also made in the late 1980s regarding the eventual establishment of a park on the Bruce Peninsula in Ontario.

The 1988 National Parks Act also contained a provision that required the minister of the environment to present to Parliament every two years a State of the Parks report. Such regular reporting was intended to indicate how well the the parks system was protecting the national heritage and also was to serve as a benchmark against which to measure progress towards completion of the national parks system. The first State of the Parks report was published in 1991 and reiterated the federal commitment, set out earlier, in the 1990 Green Plan, to establish five new parks by 1996, conclude agreements for thirteen additional parks to complete the terrestrial system by the year 2000, and complete other initiatives as well.[22]

In 1991 the Parks Service also sought to modernize the 1979 Parks policy. But the very act of opening the policy for discussion was a source of deep concern to the protectionist-oriented clientele of the service. It feared that such an exercise would simply create a political opening for those who were motivated by user concerns and tourism. Another consultation effort began in 1991 to arrive at a new consensus-based Parks policy.

The Parks Service tried to pave the way for the 1990s by enunciating its own internal strategic plan in October 1990, which was published under the signatures of all of Parks Canada senior management. Of greatest interest was the new wording through which it was hoped the 'protection versus use' value conflict could be mediated or minimized. The strategic plan spoke of three overall purposes for the Parks Service: to 'protect through science,' to 'present through experience' and to 'foster national identity.'[23] Canadians would henceforth 'experience' rather than merely 'use' the parks and hence appreciate their environment more completely and passionately. Protection would be accompanied by a commitment to use science in the parks and the building of links with others who have knowledge of how to protect the heritage involved.

It is often easy to be critical of mere changes in wording. But these latest attempts undoubtedly reflect a genuine search for values that will unite a diverse set of environmental views and interests at a time when money and resources are thinly spread but expectations are high.

Conclusions

The Parks Service was often a reluctant partner in the overall DOE enterprise. Joining the environmental club at the end of the DOE's first decade, the Parks Service accurately viewed itself to be primarily a cash cow for the rest of the beleaguered department. Some of this suspicion lessened in the latter part of the 1980s, but never totally. The Parks component had to increasingly resolve the double tension inherent within its mandate and its resource situation, one between its preservation and ever-more-diverse user goals and the other between its desire for new parks and its need to sustain the old ones. Of a lesser order of magnitude was the tension between natural parks and heritage protection.

The South Moresby decision reveals the double-edged nature of these larger tensions. On the one hand, South Moresby was a triumph of local and national environmental pressures. Such real and moral victories are undoubtedly important and show the vital necessity for outside pressure to enhance environmental values. But at the same time, each new park has the potential to rob the existing parks of resources, at least to some extent. And, as the South Moresby case showed, the great bulk of the money was needed to buy out the logging permits held by logging companies. The same environmental groups that rightfully cheer the Parks Service on when a new park is possible are ultimately, along with forestry interests, among the reasons why the Parks Service must move episodically towards its long-term plan, which many environmentalists otherwise support.

None the less, it is clear that Parks Canada does have a longer-term plan that commands a reasonable amount of political support, albeit with much grumbling about how fast the resources are being made available. The Green Plan funding of the early 1990s promises some stability of resources for Parks Canada but does not fundamentally resolve the underlying user-versus-protection goals, which always reappear in new ways.

In this realm of environmental policy, as in others, there is no escaping the ecosystem and the scientific uncertainties inherent in the otherwise simple word 'protection.' First, there is the inherent amount of land and space involved. Canada's parks exceed the size of many other countries. Second, there is the practical task of identifying the natural territorial regions that policy is seeking to preserve. Third, as the early South Morseby decision showed, there are complex linkages between logging and fish habitat, and between both and the preservation of species essential to Haida life and traditions. When recent Parks Canada studies advocate the need to 'protect through science' and to enhance the role of parks as environmental benchmarks for gene pools, it is not hard to agree. Nor can one do anything but conclude that such a scientific enterprise has barely begun.

9

Environmental Assessment and the Quest for Legislation

The evolution of the Environmental Assessment and Review Process (EARP) of the federal government is the central example of the most fundamental problem facing all environment departments – namely, how to move environmental policy towards a 'prevention' mode rather than a process of 'reaction' to pollution that has already occurred.[1] For such a move to be possible, a system of decision making involving the prior assessment of projects (and eventually policies) had to be established. The analysis of EARP presented in this chapter reveals the many difficulties that can be encountered in attempting to prevent environmental damage and in managing natural resources more effectively.

The quest to develop a legislative base for the EARP is also a central theme in the story to follow. The EARP, prior to developments in 1992, was not a legislated process, but was instead based on a Cabinet guidelines order. Each time the process was put through yet another round of scrutiny and attempted reform, officials of the Department of the Environment (DOE) attempted to push the idea of legislation. These efforts were repeatedly repulsed within the corridors of power until key court cases in the 1990s forced the government's hand.

The EARP was also tied up with everything else that was happening to the DOE throughout much of its history, including the problems of limited resources, lack of attention from the political centre, pressure from the environmental lobby, and vulnerability to the ever-present pressures for economic development by business and the provinces.

This chapter proceeds in four sections, each addressing a specific period in the evolution of the EARP. The first section discusses the build-up to the first Cabinet decision on the EARP in 1973. The second section analyses the EARP reforms embodied in the 1977 Cabinet amendments. This is followed by an account of the major reform of the EARP in the 1984 Cabinet guidelines order. And, finally, the early 1990s Federal Court cases interpreting the order are examined as a watershed in the history of the EARP and as key events which have finally brought a legislative base to reality. This final section also includes a brief mention of the concurrent effort in the 1990s to subject all government decisions, rather than just projects, to prior environmental assessment. This process too starts with a guidelines approach.

Early Caution and the Avoidance of U.S. Excess

Before looking closely at early developments, it is essential to have a basic appreciation of how the guideline-based EARP process proceeds. Every federal government department and agency, excluding crown corporations and certain regulatory agencies, has important responsibilities under the EARP process. They determine, in an 'initial assessment,' the 'environmental significance' of the projects they are initiating or sponsoring. The EARP applies to 'federal projects,' which are considered to be those that are initiated by federal departments and agencies, those for which federal funds are to be used, and those involving federal lands and jurisdiction.[2]

All processes of environmental-impact assessment require some element of initial self-assessment by departments and agencies because of the sheer number and complexity of the project decisions taken by government (about 10,000 annually in recent years). The issue then becomes one of determining to what extent the self-assessment stage is monitored centrally with regard to adherence to the letter and spirit of the process.

When a department or agency initiating or sponsoring a project deems it to have potentially significant environmental impacts, it asks the minister of the environment to establish an Environmental Assessment Panel (EAP). The panel reviews the proposal and recommends appropriate actions, which can include allowing the project, disallowing the project, or allowing the project with modifications.

The DOE's Federal Environmental Assessment and Review Office (FEARO) administers the assessment-panel stage of the process. Panel members are provided with support as they review scientific evidence and consult with experts and the general public. The FEARO then attempts to encourage integration of the panel findings into the project planning and implementation undertaken by the initiating or sponsoring department.

Prior to the formation of the EARP, however, Canada, like other Western democracies, had largely avoided taking the environment into account in project planning and implementation. The modern era of natural-resource development led to the opening of Canada's hinterlands on a massive scale. The federal government encouraged these developments through various policies, programs, and projects intended to encourage regional development, national integration, and job creation. There was an emphasis on mega-projects with very high financial, social, and environmental costs.

A long series of dams and diversions, mines and smelters, pipelines and paper mills, and chemical and nuclear plants were considered first as economic-development projects. Displacement of aboriginal peoples, destruction of animal habitat, and pollution side-effects such as mercury contamination of fish resulting from dam diversions were simply not accounted for.[3] By the early 1970s a total lack of interest in these costs became a much less feasible option.

Canada's EARP benefited from the initial flurry of government activity on the environment in the early 1970s. In May 1972, the Trudeau Cabinet authorized the Canadian delegation to the Stockholm Conference on Human Settlements to state that Canada 'considers study of the environmental effects of major development projects to be of a high priority.'[4] Earlier, on 8 January 1972, the Trudeau Cabinet noted its intent to establish in the near future a process that would subject all projects initiated by the federal government, with the exception of crown corporations and certain regulatory agencies, to screening.[5]

The first assessment process, however, had already been established in the United States some three years before the Canadian EARP was started. The National Environmental Policy Act (NEPA) was signed into law by President Nixon in 1969 following a large majority vote in both the Senate and the House of Representatives.[6] The NEPA had a tremendous impact on the development of the EARP in Canada, in large part because Canada had not yet accumulated a pool of experience on environmental assessment.

The NEPA had a number of characteristics and results which influenced Canadian EARP administrators.[7] First, the NEPA did not create a separate organization to administer the legislation, and there were no separate provisions for public participation. Accordingly, litigation became the only means for the resolution of conflicts. The U.S. courts granted standing to environmental groups in the judicial review of administrative actions. These groups made effective use of this standing to force the U.S. government to adhere to its own law regarding the preparation of environmental-impact statements or to ensure that the impact statements which were undertaken were not superficial. Not surprisingly, this led to a long series of time-consuming and expensive court cases. It also led to growing resistance from U.S. governmental agencies and successive presidential administrations.

Two further features of the U.S. assessment process were of major importance. The first feature was simply the fact that it was a legal requirement. Individuals and groups were able to use this aspect of the NEPA to stall or prevent developments that they believed were environmentally damaging. The second characteristic related to the coverage and scope of the U.S. process. Assessments not only were to consider the 'environmentally significant' effects of specific development projects, but also could cover legislative proposals and major programs. The scope of these assessments could also be comparatively broad, considering not only 'environmental effects' but related 'aesthetic, historic, cultural, economic, social, and health effects' as well. As one senior DOE official put it, '[The] NEPA was really a sustainable-development initiative long before its time.' But, not surprisingly, this admirable ambition clashed with dominant value systems and interests in government and business. This opposition, combined with the major litigation problems referred to above, provided many interested observers with an early indication that the NEPA might not be as successful as many had hoped.

DOE officials involved in developing the Canadian process were chastened by these early U.S. results. They were not at all certain about what an alternative process should be, but as one senior official noted, 'The U.S. experience was far more significant to us than any experience we had here in Canada.' The U.S. EARP was seen as indicative of what could happen when a process was entrenched in law before everything was worked through. In particular, DOE officials wanted to avoid the unpredictability and costs associated with using the courts. They sought a viable process and organization for environmental assessments, including broad public participation. The DOE strategy was to wait until the conditions for change had improved before pressing for legislation.

However, Jack Davis, as the first minister of the environment, remained very bullish on the idea of environmental assessments. He wanted to pursue the possibility of a legislative basis for the EARP even before his officials saw their way clear. Davis also pressured his Cabinet colleagues to be open to the idea. For him it was obvious that the most effective route to environmental protection was through prevention.

An interdepartmental committee of deputy ministers, convened to look at the question of the EARP, promoted the idea of a cautious approach far more vigorously than anyone else. Early assessment panels, it was argued, should be madeup of bureaucrats who would treat the assessment process more informally. The panels would conduct environmental impacts in relation to a 'statement of principles' and reach administrative compromises around those principles. As one ex–DOE official put it, 'They did not want anything too demanding ... they wanted flexibility.' For these deputy ministers, the ideal process was one that could be managed largely within the confines of their own organizations, and thus they encouraged their ministers to move with caution.

The position being suggested was revealed clearly in a 1973 memo dispatched by Gorden Osbaldeston, secretary of the Treasury Board, to Robert Shaw, deputy minister of the DOE.[8] Osbaldeston noted that, while he agreed with the DOE proposal that the EARP should, 'to the extent possible,' determine the 'environmental effects of major development projects,' he was not sure that the need for 'balance' was yet being properly addressed. On this note, Osbaldeston continued:

In this regard, we do not think the proposal adequately takes into account certain concerns raised at the meeting of the Interdepartmental Committee on the Environment, on June 29, 1973. These concerns are noted in the Summary of Discussion of the meeting but, in particular, I am referring to the general recommendation of other departments that the Environmental Assessment Panel be constituted on an interdepartmental basis.

More specifically, our view is that the 'proponent' department or agency should be intimately involved in all steps of the assessment and review process and participate in the assessment panel.[9]

Osbaldeston then went on to reveal his uneasiness with the coverage and scope of the EARP when he noted: 'I am also concerned with the open-ended nature of the proposed assessment and review process (all environmental effects of all major federal projects and activities) and the consequent effects on both the management process and the financial requirements within and without the federal government.'[10]

Retreat from the fledgling idea of a legislated EARP came in the form of the December 1973 decision by the Cabinet which created the process. The decision established that the EARP was to be based on self-assessment. Federal government departments and agencies were to develop their own screening procedures and apply them to their own proposals.

Crown corporations and key regulatory agencies would only be 'invited' to participate, and policy adopted by Cabinet would not fall under the assessment process. Projects deemed to have potentially 'significant' environmental impacts were to be referred to the minister of the environment for review, but the assessment panel was to comprise only bureaucrats from the DOE and the initiating department. Public reviews would not yet have a major role to play, and socio-economic factors were not to be part of environmental-impact assessments.

Cabinet ministers had obviously agreed with the argument that, because little was known about environmental assessments, caution should be the order of the day. The EARP was to be left suitably vague and was not to be overly rigorous in terms of monitoring and follow-up, so that it could be gradually worked through the power politics of federal-provincial relations and the federal bureaucracy. The coverage and scope of the process should also not be such that the government's major development initiatives could be 'unnecessarily' detained through bureaucratic red tape. A stock of experience at some later date might allow for formal legislation with fewer unknown qualities and quantities.

Within this framework, Jack Davis sought to use the Lands Directorate as the vehicle for his early assessments. As we noted in chapter 1, the Lands Directorate brought together a considerable base of knowledge regarding land-use planning and the relations between the environmental media of land and water. It already had the Lands Inventory, a key information source, and a solid tradition of cooperation with industry and the provinces. This cooperation was seen as essential to a successful EARP.

From very early on, however, it was evident that the Lands Directorate was not equipped to do the job. It was a headquarters operation with no capacity in the regions, and, as it found out, it had insufficient technical know-how relative to the complexity of the ecological issues being faced. The Lands Directorate sought to compensate for these limitations by learning as much as possible from the U.S. experience. But, as one senior official noted, 'There was a realization that the existing departmental policy and approach [were] just not sufficient.' Jack Davis also began to lose some of his enthusiasm for the EARP when the first major report by the Lands Directorate regarding a second bridge over Vancouver's Burrard Inlet recommended that the project not proceed. Davis buried the report because the project was simply too sensitive in his own political riding in British Columbia. Similar problems would repeatedly arise to be faced by the DOE as it sought to insert its assessment agenda into the mainstream of government.

The Lands Directorate was replaced as the administrator of the EARP by the FEARO in April 1974. The FEARO was intended as a quasi-independent organizational component within the DOE.

The FEARO would not be any easy answer to EARP problems. First, because no mention had been made of an administrative body for the EARP in the 1973 Cabinet decision, the FEARO was left without a solid base of authority. Second, the concept of 'self-assessment' left the organization in a weak position. Add to this a lack of financial and staff resources, and it is not difficult to see why the FEARO was basically a weak secretariat prevented from doing anything more than providing basic administrative-support duties for assessment panels.

Reed Logie, the first chairman of the FEARO, outlined in a memo to Robert Shaw the resources that the new organization would need.

I do not want a panel with a large fixed membership. I think I need for the Panel:

1) an energy expert, who thinks environmentally;
2) a construction engineer, who also thinks environmentally;
3) a biologist, who thinks pragmatically.

I also need on my staff, but not necessarily on the Panel, a liaison officer, to establish and to maintain the many contacts with the other departments and, where required, with the provinces and with environmental assessment project managers. All these people must be top notch personally and professionally, because the Panel will live or die by the credibility that it creates for itself by its performance.[11]

The FEARO was clearly not going to be a major force for a new environmental-policy thrust. A decline in the EARP's credibility was not long in coming.

In fact, it was not much more then a year later that problems with the EARP and the FEARO began to appear with greater frequency. The extent and character of these problems are captured well by two 1975 memos sent to Blair Seaborn, the then-new deputy minister. In the first of these memos, Ghislain M. Gauthier, chairman of the Quebec Council of Regional Directors, informed Seaborn that

in the Quebec Region, EARP has until now, benefitted from favourable prejudice in the other federal and provincial ministries. The time has come to show more "leadership" and in certain cases more firmness.

This would require our regional officers to play a more dominant role. Following the flutter created by certain projects in the region, such as James Bay and the Sarnia-Montreal Pipeline, the public expects the governments to proceed with public hearings in the course of construction of large projects.

The provincial officers seem well disposed to act, but require active moral and technical assistance from our Ministry. It is essential that our officers be assured administrative support required to this end.[12]

By this time it had become very clear that the resources of the FEARO were not nearly up to the task of providing this support. Further, with self-assessment proceeding without monitoring, and very few avenues for public involvement, the credibility of the EARP was lagging badly.

In the second memo, André A. Grignon, director of the Federal and Provincial Programs Branch, expressed his concerns.[13] He specifically addressed the failure of the EARP to cover crown corporations and regulatory agencies and noted:

It is now obvious that only 'inviting' Proprietary Corporations and Regulatory Agencies to participate in the EARP is insufficient. As indicated in the attached document, a large number of those Corporations and Agencies are precisely those which should be involved in EARP on a priority basis. Examples are those Proprietary Corporations involved in transportation by land, air and sea, and Regulatory Agencies such as the National Energy Board and the Canadian Transportation Commission.[14]

It was evident that it was not enough to simply 'invite' organizations to participate. Discretion by these agencies would clearly not be surrendered that easily.

Public Involvement and the 1977 Reforms

Environmental groups were extremely critical of the early assessment process, and thus by the mid-1970s there was a desire within the DOE to seek reforms. The extent and nature of the DOE's concerns were brought out most clearly in a discussion document sent by Blair Seaborn to the members of the Interdepartmental Committee on the Environment on 14 January 1976.[15] One of the chief concerns expressed by Seaborn was the limited extent of public involvement in the process and the role this involvement could play in improving the accountability of the EARP. By the mid-1970s, in part through the dynamics of the Berger Commission, the DOE was becoming more aware of the role that public involvement could play in mobilizing support for its agenda. Exposing the process to public scrutiny was the most likely way to ensure its integrity and greater accountability at the initial self-assessment stage.

Seaborn also wanted to strengthen the monitoring and follow-up of the process and enhance its accountability. As far as the DOE was concerned there was no way to assess how well the EARP was doing because only those projects deemed 'environmentally significant' were referred to the DOE. What proportion these constituted of the total projects being initiated or sponsored could not be assessed. No one knew how many projects were entering the process or how many 'significant' ones were avoiding the panel-review stage. He recommended the provision of information upon which the process could be judged in public forums.

Finally, Seaborn was interested in delineating more clearly the responsibilities of the DOE and the FEARO in the process. Specifically, he addressed the financial responsibilities of the FEARO. One of the major constraints on the process had proven to be the limited resources available for FEARO to initiate reviews, provide for monitoring and follow-up, and inform the public. In response to this problem, Seaborn sought to reinforce the 'polluter pays' principle in the EARP whereby the initiator or sponsor of the program should be responsible for covering costs associated with complying with the process. He was primarily concerned with allocating costs to non-government proponents. Accordingly, he recommended a schedule of cost-related responsibilities. By doing this, it was hoped that one substantial barrier to the process could be overcome.

However, by the mid-1970s it was clear that the conditions faced by the DOE in the second half of the 1970s were not going to be the same as those faced in the first half of that decade. Seaborn's proposals for EARP amendments hit the Ottawa policy network in the period when severe structural problems in the economy were becoming very evident. The Nixon Shocks of 1971–2, the first oil crisis of 1973, and 'stagflation,' a combination of high unemployment with rising prices, hit hard in the mid-1970s. Accordingly, the federal government's priority list was dominated by issues of energy policy, inflation, foreign ownership, manufacturing competitiveness, science and technology, trade relations, and regional development. The 'obscure' EARP process was definitely well down the list. Legislation for the EARP, receiving some discussion within the DOE, was to be, once again, a non-starter.

The extent to which DOE officials experienced barriers to the EARP could be seen in a memo from Blair Seaborn to Romeo LeBlanc on 12 November 1976.[16] In that memo Seaborn impressed upon the minister that, while he also wanted to 'develop a new approach based on legislation,' the time was not right. LeBlanc had expressed his concern over the 'self-assessment approach in our present process,' but Seaborn had to conclude that the best he could get was a 'memorandum to Cabinet recommending changes to the existing environmental assessment process.'[17] It was the only option that would gather the needed support from other federal departments and agencies. On this note he made his case to the minister,

It has taken a considerable amount of time to develop the procedures for, and promote an understanding of, the existing process since it deals with very complex questions. A number of important projects have just recently been referred to the Environmental Assessment Panel. It will take some time to determine how adequately the process can deal with these. The experience gained over the next months should significantly add to the information base needed to assess the process as well as establish the need for new approaches. The amendments suggested in the memorandum to Cabinet will facilitate this process appraisal.[18]

The DOE was still on a steep learning curve and was sensitive to how far it could push the EARP in the corridors of power in Ottawa.

The post-1975 period, as chapter 4 showed, was also a period when the DOE was delegating much of its environmental compliance and enforcement activities to the provinces. The federal government was seeking provincial cooperation with its economic-development initiatives and did not want any roadblocks to prevent this new partnership from flourishing.

In the late 1970s, several provinces were also actively developing their own environmental-assessment processes. Much of this activity can be explained as a pre-emptive anticipation of federal actions impacting on provincial jurisdiction. By 1978, Quebec had established its EIA process of environmental-impact assessment through an amendment to its Environmental Quality Act. Throughout the 1980s, all the other provinces would see their efforts to develop environmental assessments come to fruition. Along with these provincial initiatives came a much greater likelihood of conflict between the federal and provincial governments over such assessments. It also meant that those opposing the federal EARP process could point to the provinces and say 'let them handle it.'[19]

The difficulties faced by the DOE were reflected in the modest EARP reforms that were adopted by Cabinet on 8 February 1977. The FEARO was allowed to recruit assessment-panel members from outside government, but, as with the original 1973 decision, there was no mention of an administrative body with clear authority to coordinate and monitor the process. One DOE document put it this way: 'In effect, this amendment limited itself to modifications of the panel composition and stated that existing review mechanisms should be strengthened without specifying how this should be done.'[20] Self-assessment was confirmed as the basic principle of the EARP, and public involvement was not going to play much of a role in monitoring its implementation. One senior DOE official noted, 'It did not take long before we began to notice that not much had changed. The pre–panel assessment stage of the process had not adequately been addressed. Projects were still able to slip through the cracks.' Almost immediately another round of efforts towards reform had to begin.

One new thrust for reform came in the form of the 1979 Government Organization Act, which separated Fisheries from the DOE. The act gave the DOE some horizontal coordinative powers within the federal government and is therefore an important development in the EARP story, although one rarely given much attention. It stated that the minister of the environment

shall initiate, recommend and undertake programs and co-ordinate programs of the Government of Canada, that are designed ... to ensure that new federal projects, programs and activities are assessed early in the planning process for potential effects on the quality of the natural environment and that a further review is carried out of those projects, programs, and activities that are found to have probable significant adverse effects, and the results therefore taken into account.[21]

After the 1973 and 1977 Cabinet decisions, the DOE's role in the EARP was generally recognized, but its efforts to monitor the process had been made virtually impossible because it could not appeal to any specific horizontal powers. Other government departments and agencies were able to claim that the EARP was the equal responsibility of them all. While this is no doubt the ideal in theory, in reality it never works that way. The final responsibility for any policy process must reside somewhere, and early Cabinet decisions did not help clear the air on this issue. Other departments and agencies were less committed to the environmental issue and consistently argued they were meeting their obligations in-house and would appeal to the DOE only for its technical expertise.

The origins of the new DOE powers can be traced in part to manoeuvres by DOE officials and in part to early discussions between Prime Minister Trudeau and Leonard Marchand, Minister of State for the Environment. In one letter sent by Trudeau to Marchand on 9 November 1978, the prime minister expressed his support for the environment department and stated,

I believe the Government Organization Act should empower the Minister of the Environment, with the approval of the Governor in Council, to set environmental guidelines for departments to follow in drafting legislation, setting regulations, issuing permits and other program activities. It would be the responsibility of the Minister of the Environment to advise his colleagues on the environmental questions and on the implementation of the environmental guidelines.[22]

On the specific question of the EARP within these horizontal powers, Trudeau continued:

At the same time those guidelines should inform the Environmental Assessment and Review Process (which should be given a legislative base patterned on its current procedures and methods of operation). A separate organization to administer that Process on behalf of the Minister also should be established in legislation.[23]

All of these new powers, however, were to be severely constrained by the Trudeau government's economic and federal-provincial policies. We have already seen, in chapter 2, how Trudeau's legacy on the environment could be viewed as initial support followed in short order by retrenchment and lack of interest in the face of economic and constitutional concerns. Thus, again in this case, Trudeau explicitly ordered that

in exercising your new horizontal powers, and in particular in developing the Environmental Assessment and Review Process, you will need to ensure that the objectives of the Government to reduce the regulatory burden on the private sector is respected and emphasized. In its operation to date EARP has demonstrated it can be effective while remaining small in size and avoiding red tape and delay. You should build in safeguards to the new system to ensure that it continues to operate in this manner and does not succumb to bureaucratic tendencies to expand for the sake of expansion or to create a regulatory behemoth.[24]

At the same time that Trudeau was promoting horizontal powers for the DOE he was willing to enshrine many of the past problems of the process. He did not want the EARP becoming 'regulatory' and a 'burden' on the economy, and his concern for federal-provincial peace was revealed in his call for a delegation of DOE programs 'to the provinces.' However valid many of these concerns may have been at the time, they served as powerful rallying points for other departments and agencies which sought to resist the interference of the EARP in their discretionary activities. His call for horizontal powers was a step forward, but in large part he simultaneously pre-empted any early gains that could be realized from them.

Another wave of efforts at EARP reform had to await the results of the 1979 federal election. The arrival of John Fraser in the minister's chair following a Conservative victory was a pleasant surprise for DOE officials and for environmental groups. Among the policy areas that Fraser sought to move on was the EARP. He kept hopes for a legislated EARP alive within the DOE, which had impacts beyond his short tenure.

However, within a few months, the arrival of John Roberts as minister of the DOE meant that the reform of the EARP was delayed again. Roberts took on other issues as his main priorities, not the least of which were the Liberal constitutional-reform initiatives. However, the work of DOE officials on the EARP continued. In particular, Roberts's tenure saw the conduct of yet another interdepartmental review of the EARP, which was to culminate in recommendations to Cabinet.

The Guidelines Order and Its Aftermath

By mid-1983 the EARP consultations under John Roberts's tenure were, for the most part, complete. However, once again political and economic conditions for reform were not propitious. By 1983 Canada was still reeling from the worst recession since the 1930s, and economic issues were again dominant. It was in this context that Raymond Robinson, appointed executive chairman of the FEARO in early 1982, provided Roberts with his recommendations regarding the EARP.[25] According to

Robinson, the point had been reached where definitive statements could be made about the positions taken by 'senior officials in the departments consulted.' He then went on to review these positions and provide the minister with several issues which needed his attention and advice, not the least of which was, once again, the question of legislation.

Robinson addressed many of the reform issues that had been adopted by Seaborn eight years earlier, plus others that evolved since. Pressured by the ENGOs, but also moved by its own dissatisfaction, the DOE was still seeking to tighten the initial assessment stage, broaden the scope of public reviews, improve monitoring and follow-up, widen the coverage, and deal with questions of funding the process. On most of these issues progress over the first decade had been marginal, and Robinson informed the minister that, 'as a general observation, I have to conclude that most of the senior officials prefer retaining as much flexibility as possible in any new regime. In particular EMR and Transport are not persuaded that any basic change is needed although they naturally support our proposed tightening of the terms of reference for panel reviews and greater administrative control of panels.'[26]

In general, Robinson had to conclude that on each of the reform issues identified by the DOE as priorities, other departments were not going to volunteer their support. On the question of initial assessment, he noted that 'EMR and DOT believe the present system is adequate and requires no real change.' Similarly, on the scope of the process, it was concluded that many of the major departments of government did not want the EARP to consider more than the 'biophysical and directly related social impacts.' Coverage should continue to exempt 'Crown Corporations and regulatory agencies,' and there was 'less enthusiasm for a post-assessment role for Environment Canada as overseer of the implementation of panel reports.' On the issue of public participation, the findings were equally disappointing. Robinson concluded that many departments 'do not like the suggestion that they be required to undertake public consultation during the screening process. They want discretion in this matter.' Similarly, the 'intervener funding' of public participation drew 'concerns about the allocation and administration of funding' and arguments about the 'need to follow tight procedures and to establish demanding criteria.' Robinson summed up his observations by saying,

A major sticking point is any attempt to introduce a review mechanism, especially one open to receive complaints from the public, to second guess the judgement of initiating departments on whether a project should be referred to you for public review. As noted above, there is even opposition on the part of Transport and EMR to a requirement for public consultation during the initial screening process although regular publication, after the event, of initial screening results would be acceptable ... I am morally certain that continued insistence on this point would provoke strong opposition from EMR, probably supported by Transport.[27]

On the question of EARP legislation, the situation was even more bleak. Robinson was not convinced, based on the experience of his first interdepartmental EARP battle, that the DOE should even attempt to proceed with a legislative agenda. The best he could do was point to departments that were 'least hostile' rather than to those who were supportive of the idea. He was sure that, under the conditions that dominated the Ottawa system at that time, the DOE could never demonstrate the 'need for legislation to senior officials from major departments.' Instead he wanted to know the minister's view 'on whether the Cabinet Submission should maintain our present emphasis on the legislative option, with an Order-in-Council as a fall-back, or whether the paper should highlight the Order-in-Council approach with legislation as an option for the record.'[28]

It did not take long to see that the legislative option was going to remain only 'for the record,' at least for the foreseeable future. Less than two months later John Roberts was no longer minister, and concerns over the state of the economy remained paramount. In fact, any further movement at all on the EARP had to await the appointment of Charles Caccia as environment minister, and the run-up to the 1984 general election.

Charles Caccia took over as minister in 1983 well aware that he had only one year to work with. One of the areas he set to work on was the EARP, and he found that some ministers were beginning to awaken to the fact that the environment was becoming a major issue for the voting public. Caccia accepted the general DOE assessment that legislation was not a realistic option, particularly in the time he had at his disposal. While there was some softening of views within the Trudeau government, he did not, as was the case for his predecessors, have the political muscle within Cabinet to force the EARP legislation over the objections of senior bureaucrats and intransigent ministers. According to him, one of the main problems faced by the EARP was that the process, after some ten years, remained largely unwritten and vague, and he wanted to have it codified as 'an administrative code of practice.' The Robinson recommendation to proceed with an order-in-council submission to Cabinet was a compromise that went some distance to meeting Caccia's main concern while holding out some prospect for success in a new round of EARP reforms.

The guidelines order was passed by Cabinet in June 1984, but not until after a Laurel and Hardy–like event which would later prove to be significant in the evolution of the EARP.[29] As one senior DOE official noted, 'We thought we were just codifying something that was already happening within government informally, but ended up getting a lot more.' This result is attributable to the rush to get the guidelines order to the last meeting of the Trudeau Cabinet in 1984 before a new Turner Liberal Cabinet took office.

The DOE sent a rough draft of the order to the Department of Justice within what many in the FEARO thought was adequate time to put it into formal language. However, after a series of sloppy drafts were returned to the DOE, confusion arose as to how the process should be represented in written text. This led to a situation whereby, as one senior DOE official put it, 'there was no time left to work it out. The result was an order with a character and sound which went beyond even our original proposal!' Another official added, 'We thought it was going to have good administrative force, much like a Treasury Board policy directive. It would be an advance simply because we had so little before, but it ended up taking on a broader legal meaning.' As we will see, it took some four years for the real extent of the 'legal meaning' to become clear to all concerned.

The order proved to be a significant step in the evolution of the EARP for at least three important reasons in addition to the basic fact that it formally 'codified' the process, and identified the FEARO as the agency 'responsible directly to the Minister for the administration of the process.' First, it established that the EARP applied to 'any initiative, undertaking or activity for which the Government of Canada has a decision making responsibility.' It required that each initiative 'shall be subject to an environmental screening or initial assessment to determine whether, and the extent to which, there may be any potentially adverse environmental effects from a proposal.'[30]

Section 12 of the order stated that a number of results can follow from the initial assessment stage, including the proposal being allowed to 'automatically proceed'; the proposal being 'referred to the Minister for public review by a Panel' because it has 'potentially adverse environmental effects'; the proposal being allowed to proceed with mitigating steps taken with 'known technology'; the proposal being subject to further study owing to 'unknown' environmental effects; and the proposal being 'modified' or 'abandoned' because of 'adverse environmental effects.' However, even if a determination is made by government officials that the environmental effects are 'insignificant' or not 'adverse,' Section 13 contemplates a review being undertaken if 'public concern about the proposal is such that a public review is desirable.' This would later prove to be a very important addition.

Second, if a public review is considered necessary, Section 22 requires that the assessment panel shall meet three criteria: first, that it be 'unbiased and free of any potential conflict of interest relative to the proposal under review'; second, that it be 'free of any political influence'; and third, that it 'have special knowledge and experience relevant to anticipated technical, environmental, and social effects of the proposal under review.' These criteria meant a broadening of public access to the panel-review stage, and a process less directly controllable within the confines of specific government departments.

Third, the order requires that the minister issue the terms of reference for panel reviews to the public, and that these terms shall include 'an examination of the environmental effects of the proposal' and 'an examination of the directly related social impacts of those effects.' This gave some room for increasing the scope of the process. Further, with the approval of the minister and the initiating department, the review may also consider 'general socio-economic effects' and a 'technology assessment of the need for the proposal.' During the consideration of these matters, the panel shall undertake 'public hearings' in a 'non-judicial and informal but structured manner' where witnesses can be 'questioned.' The review panel shall also 'conduct a public information program to advise the public of its review and to ensure that the public has access to all relevant information that any member of the public may request.' This amounted to public involvement and scrutiny being given a substantial push forward with some discernible effects on the balance of power over EARP issues within the federal government and in federal-provincial relations.

However, the impact of the guidelines order, as noted above, was not to be immediately experienced in its full force. The 1984 general election saw Charles Caccia removed from the minister's chair after the Mulroney Conservative victory. The Conservative government's attempts to promote the reform of the EARP received some momentum from four politically significant events. First, the 1984 Conservative party election platform contained a commitment that there should be EARP legislation. This aspect of the party's platform was encouraged by John Fraser, who remained active on environmental-policy matters after his tenure as minister. The Mulroney government had to face this commitment every time they were challenged by environmental groups to act on the EARP.

Second, the recommendation by the Royal Commission on the Economic Union and Development Prospects for Canada (the Macdonald Commission) that the EARP be given a statutory basis added additional momentum.[31] The commission argued that environmental values needed to be better integrated into economic decisions, and that provisions for public input, intervener funding, and monitoring and compliance could improve the legitimacy of the assessment process.

A third influence was that there was an early realization by some government officials that the guidelines order might in fact have more legal weight than was at first thought. In anticipation of the possibility of strict court rulings on the order, there were calls in the DOE for a better balancing of ministerial discretion with public and judicial scrutiny of the process, and for more flexible mechanisms for federal-provincial cooperation on assessments. Legislation, if it could build some of these safeguards in, did not appear as worthy of maximum resistance as it once had.

Finally, the work of the World Commission on Environment and Development, the Brundtland Commission, added further impetus. Through the concept of 'sustainable development,' it called for better resource-management and environmental-protection measures in all governments around the world.

Even with these events, however, the road to EARP legislation remained rough. As one ex–senior DOE official noted, 'MacMillan wanted to go ahead immediately with draft legislation but could only get a discussion paper out of Cabinet. It was months of work before we could get the discussion paper made public.' By 1987 many of the key concepts regarding what the DOE wanted the EARP legislation to look like were in place. The Canadian Environmental Advisory Council also produced its review of the EARP. It advised the minister to proceed with a legislated process, but the 1988 election put everything on hold once again.[32]

The Conservative victory in the election of 1988 led to the appointment of Lucien Bouchard as minister of the environment. Chapter 2 stressed that Bouchard appeared as an aggressive minister, willing to take some risks in promoting the DOE's environmental-policy agenda, but it was not his political clout that had the most effect on the EARP story in the early 1990s.

The Rafferty-Alameda *and* Oldman River *Court Cases and Bill C-13*

Soon after the election of the Conservatives for a second term, a series of Federal Court rulings on the EARP provided the vital catalyst for a legislated process. A brief summary of the two most important court rulings, occurring in 1989 and 1990, serves as an essential background to EARP legislative initiatives in the early 1990s.[33]

The first important court ruling concerned a case launched by the Canadian Wildlife Federation (CWF) opposing the Rafferty-Alameda dams project on the Souris River in Saskatchewan.[34] The CWF sought to quash a federal licence for construction of the dams issued under the International River Improvements Act on the grounds that the minister of the environment had not complied with the federal government's own EARP guidelines order before issuing the licence. On 10 April 1989, Mr Justice Bud Cullen ruled that the project must be subject to an environmental-impact assessment. The ruling affirmed that the federal government must fulfil its obligations under the guidelines order whenever it had a regulatory responsibility, even though the dams were a provincial project, primarily involving provincial jurisdiction over natural resources.

The second important court ruling was a Federal Court of Appeal decision on a case launched by the Friends of the Oldman River Society opposing the Three Rivers Dam on Alberta's Oldman River.[35] This case involved the authority of the minister of transport under the Navigable Water Protection Act and the minister of fisheries and oceans under the Fisheries Act. A lower court had ruled that both ministers lacked authority to require an environmental-impact assessment. The March 1990 decision of the Federal Court of Appeal ruled, however, that both ministers were in fact bound by the federal EARP guidelines order. The decision thus broadened the Rafferty-Alameda decision because it ruled that the federal government had to apply the order to any initiative impacting upon federal jurisdiction even though no request for a federal licence was made or might be required.

A later appeal of the *Oldman River* case to the Supreme Court of Canada narrowed this degree of application of the guidelines order. The January 1992 decision of the Supreme Court ruled that the guidelines applied only to cases where a federal department has an 'affirmative' regulatory responsibility or 'duty.'[36]

These court cases in essence made the guidelines order a binding legal requirement, albeit, in the cases at hand, neither of the dams sponsors faced a decision to halt construction. One confidential DOE document argued that the rulings were 'forcing decision makers to comply with the letter of the Order, to be clear in the articulation of their decisions, and to ensure that procedures are public.'[37]

The uncertainty created by these cases, coupled with the fear of more court cases to come, not to mention costly federal-provincial conflict, gave the idea of EARP legislation a new lease on life. Moreover, resistance to legislation within the federal government was substantially curtailed by the court cases because the rulings enforced a strict reading of the guidelines order, thereby decreasing the extent of administrative discretion involved in its application. Growing public concern over the environment also aided the DOE's renewed promotion of legislation for the EARP.

On 8 June 1990, the federal government introduced Bill C-78, The Canadian Environmental Assessment Act (CEAA), which later became Bill C-13. The legislation was given third reading on 19 March 1992, but its proclamation has since been delayed, pending the development of its supporting regulations. Opposition to the law's regulations has come from Ottawa's economic departments and their industrial clientele, in part because such interests believe that many of the discretionary features are being reintroduced through the regulations.

While the early 1990s ended the DOE's rocky search for a legislative base and are even bringing the prospect of non-legislative prior environmental assessment of policies rather than just projects, the journey ahead is hardly an easy one.[38] The assessment of policies is a vital area but is far more radical in its impact than is the mere assessment of projects. There are serious practical problems involved in defining just what policies are at the Cabinet and departmental level and in how the ecological and scientific analytical basis of such decisions will be supplied at the centre of the Cabinet and in central decisions within departments. In addition to considerable federal-provincial and business–ENGO tension, over the content and publication of such assessments of policies, there will be an extraordinary pressure on the FEARO and its successor bodies for vastly expanded technical and analytical resources as the assessment process takes on a vastly expanded realm of activity.

Conclusions

The DOE, the FEARO, and the ENGO advocates of environmental assessment can point to some slow grinding progress on the long journey towards a legislated environmental-assessment process. While the DOE was initially determined to avoid the excesses of the U.S. legislative model, its key internal champions of the assessment process subsequently launched many efforts to lessen the wide discretion that allowed departments and interests to escape assessment. In short, they sought to eliminate the loopholes in the assessment process.

The process gradually became a deeper, broader, more independent, and more participative one, as practical lessons were learned and some areas of resistance were overcome. It took everything from accidental and rushed drafting errors to tenacious advocacy by ENGOs, and from steady pressure by some DOE officials to ENGO-led court cases, to finally reach the goal of legislative change.

But severe physical and technical constraints remain. Tens of thousands of projects and policies must in principle be assessed annually, primarily by agencies whose leaders do not put environmental matters first.

10

Legislative Catch-up:
From Fish to Toxics to the
Canadian Environmental Protection Act

If environmental assessment involved a twenty-year search for its first legislative base, then for the larger problem of environmental regulation, the dilemma was how to play legislative catch-up. A statutory base existed from the beginning for regulatory matters, but it was never good enough relative to the expanding tasks at hand. Thus our final environmental journey is a regulatory one which stretches from fish to toxics, to the promulgation of the Canadian Environmental Protection Act (CEPA) in 1988. But it is simultaneously an account of the growing environmental uncertainties posed by toxic substances and of the inherent limits of political and scientific capacity relative to the regulatory challenge at hand.

We look at the long, effectively twenty-year, CEPA journey in five phases. First, we build on chapter 4's account by delving further into the Environmental Protection Service's (EPS) early regulatory initiatives and the first political battles with industry and the provinces. Second, we look at the promulgation of the Environmental Contaminants Act (ECA) as an initial response to the new toxic contaminants and of the subsequent failure to give it adequate resources for implementation. The third section of the chapter examines early pollution-control setbacks and the intense federal-provincial conflict surrounding the 1977 Fisheries Act reforms, and a subsequent decline in EPS regulatory activities. Fourth, we trace the impact of growing public concern over toxic contaminants, and the processes that led to the development of the CEPA. Finally, we look at the implementation problems of CEPA, including equivalency agreements and attempts to improve the organizational capacity of the EPS as the Department of the Environment's main environmental regulator.

Traditional Regulation and Early Agenda Choices

Traditional environmental regulation first evolved around what one former senior DOE official called 'the first generation industrial pollution problems such as sewage, stack gas, and mercury and lead.'[1] There was general acceptance throughout the liberal democracies that government needed to defend the public from pollution that had definite public-health implications.[2] Officials looked for the quickest means to achieve needed improvements in drinking water, air quality, and the 'quality of fish as food.'

EPS environmental regulation, as chapters 1 and 4 have shown, became technology-based. This approach seeks to develop standards in accordance with the 'best practical technology' available to limit the amount of pollution that escapes into the environment.[3] In theory, only after all technologically practical steps are taken will any environmental damage be accepted. In reality, as we will see, the situation became more complex than that.

Early environmental-protection legislation was written as a blunt prohibition on all polluting emissions. After the 1960s, total prohibitions began to be replaced by legislation which allowed for 'control regulatory regimes.'[4] Essentially, this was the legal recognition that some level of pollution was bound to occur from all human production and consumption. The realistic goal was to set standards and 'control' this pollution within technological limits and within the capabilities of the natural environment to cleanse the pollution from its systems.

Prosecutions, however, were still to be associated with environmental statutes and regulations. But, in Canada, industries which have breached environmental standards have rarely experienced legal sanctions.[5] Determining what the environmental problems are, responding with 'best practicable technology,' setting standards in accordance with technological capabilities, monitoring emission levels, and prosecuting for non-compliance have never been simple step-by-step procedures.

The complexity of the regulatory process has meant that traditional environmental regulation has functioned around 'compliance agreements' of various forms between industry and EPS officials. The EPS has been compelled towards a great deal of flexibility in these agreements for several reasons: first, the growing number and complexity of pollution problems placed tremendous demands on the EPS; second, structured industrial processes and interest pressure led to concerns about employment and investment if standards were rigidly enforced rather than negotiated over time; and, third, a lack of adequate EPS organizational capacity limited what could be done to enforce standards and address new problems.

In the early period, the DOE moved on pollution through a three-pronged legislative mandate: the Fisheries Act, the Clean Water Act, and the Clean Air Act. As chapter 4 noted, the Fisheries Act was the backbone of the DOE. The 1970 amendments to the act allowed the federal government to move on pollution through regulatory control regimes rather than attempting to enforce blanket prohibitions, and the early period became dominated by water issues.

This focus on water was encouraged by the Fisheries Service and early senior management of the DOE. But, by as early as 1965, the International Joint Commission was already urging the Canadian and U.S. governments to act on sewage in the Great Lakes.[6] Moreover, by 1969, the degradation of Lake Erie had become very advanced, and the Cuyahoga River, running through Cleveland Ohio, literally caught fire, owing to the fact that it was running chocolate brown with oil and other flammable industrial wastes, sending a chill through all observers. Also in 1969, the Fisheries Service had to ban fishing in Placentia Bay, Newfoundland, and in March 1970, commercial fishing in Lake St Clair and tributary rivers was halted because of mercury pollution. Jack Davis indicated his concerns about these frightening trends in the Commons Debates of 20 April 1970, when he stated, 'Anything that harms fish ... may be harmful to man himself. It follows that the living resources in water are our first line of defence. Healthy fish mean a healthy environment and a healthy fishery is undoubtedly the best insurance policy we can buy in our battle against pollution in water.'[7]

The Clean Water Act, promulgated in June 1970, was also a media-based piece of legislation, enabling federal-provincial cooperation over water-management issues. Part I of the act provided for general federal-provincial cooperation in studying water management and quality problems. Part II provided for the establishment of a water-quality management agency to manage and regulate the use of water in any particular water basin. Part III provided for control of nutrients such as phosphate in water through product controls on detergents.

The principles of Part I and the regulations under Part III were the only two parts of the Clean Water Act ever used. However, as we saw in chapter 3, the enabling aspects of the act became much more difficult to use within federal-provincial relations as budgets dried up after 1975.

The Clean Air Act, which received royal assent in June 1971, was the third media-based enabling piece of legislation. However, because it dealt with air pollution, it was not surrounded with the same degree of controversy as the Fisheries Act or the Canada Water Act. In the early period, as chapter 4 showed, air issues were not seen to be as severe or complex as many water-pollution problems. Moreover, a tradition of federal-provincial cooperation on air issues had been established by those components of the DOE which had come over from the Department of Health and Welfare. Thus there was general agreement that the federal government was best positioned to set national air-quality standards in the defence of public health, while the provinces could be the most effective at implementing them. The result was the successful completion of several regulations dealing with such matters as lead in gasoline, vinyl chloride, chlor-alkali mercury, and asbestos mines and mills.

But, it was abundantly clear that what was to occupy most of the EPS's time in the early period was water pollution and the Fisheries Act. Jack Davis, as minister, was very concerned with Fisheries issues, and he was acutely aware that a majority of the DOE's capacities as environmental regulator resided on the water side. In fact, after his statement in the House of Commons on 20 April 1970, it became evident that this was also the position of the Conservative Opposition Environment critic, who stated, 'According to the Minister of Fisheries the Fisheries Act, which we are presently amending, is all that is required to reduce, control and police pollution in all the waters of Canada. This is a viewpoint with which I agree ...'[8]

Davis also indicated an intention to be very tough on polluting industries. In a speech to the House of Commons on 15 October 1970, entitled 'Shaping our Environment in the 70s,' Davis began with:

There is real urgency here. We have to act on the environmental front quickly and with determination. We have to move ahead of events, rather than from crisis to crisis. Our critical path, in other words, must be laid out ahead of time. We must have a plan to preserve the quality of our environment. We must preserve our wildlife and our fish and our trees. We must renew our renewable resources as quickly and as effectively as we know how ... Industry, therefore, must keep its poisons to itself. Canada's cities and towns must do likewise.[9]

Davis also revealed what would be one of his earliest regulatory initiatives through the EPS when he said:

Take the liquid metal mercury for example. In its elemental form it is known in nature. But man has stripped it of its sulphur. He has released the metallic mercury and shipped it out to industry to use as it wishes.

Industry, in turn, has used elemental mercury in an indiscriminate manner. It has been using mercury like it was going out of style. It has been using it up like a modern raw material. It has been letting metallic mercury loose on our environment without thinking about its effects on our forests, on our wildlife, on our fish, and worse still, at the top of the food chain — on man himself.[10]

But how to proceed with regulation was not a clear-cut issue. As chapter 1 emphasized, the EPS was largely a headquarters organization, which, because it was new, needed to spend considerable time orienting itself to its new mandate. It had to discover what the range of pollution issues were throughout the regions of the country, and what the priorities, under conditions of limited resources, should be. The strategy adopted within the organization was to set up industry–government task forces to determine the nature of the problems. As one senior official noted, 'When we consulted on regulations, mainly only industry was involved ... until 1981.'

The selection of industries to regulate was not a particularly scientific enterprise, because it had to be based on a certain realism. One official who was involved in the process stated, 'It was very difficult for us to organize our early response because there was so much going on.' Another noted, 'We decided to concentrate on the big industries that used some sort of chemical process.' Priorities were to be set by making an assessment of the ecological seriousness of the problem and then an informed judgment about the likelihood of success in addressing it. There was no use consuming limited time and resources on losing propositions.

Judging whether a pollution problem was a 'losing proposition' or not was based on many factors: first, whether or not the media and public were providing the problem with a high-enough profile to mobilize a response; second, whether the industrial and provincial lobby was simply too strong to overcome; third, whether the technology of the industrial process was too complex; and fourth, whether the DOE simply did not have the time, money, or jurisdiction to act on that specific problem.

One senior EPS official cited three examples of early decisions: organic chemicals, municipal sewage discharges, and the steel industry. The EPS's conclusions were that, in organic chemicals, 'we did not develop regulations because of the tremendous technical difficulties.' In the area of municipal sewage, 'when we tried to move on large industries linked to municipal sewage systems, the battles with the provinces really began.' In the steel industry, 'we backed off because it was all in one provincial jurisdiction, Ontario.' Industry and a few key provinces united to make the case for 'jobs' and the jurisdiction of the provinces. Ontario, by contrast, increasingly pursued a strategy of competing with the federal government on environmental regulatory issues.

It should be recalled from chapters 1 and 4 that the organization of the EPS was itself media-based. The EPS had air, water, and land sections, each concentrating on specific pollution problems. By June 1983, the EPS would be reorganized on an industry basis by Bob Slater, its assistant deputy minister. But it was the water people who consumed the lion's share of limited organizational resources throughout the early period, and it was very difficult for them to accomplish goals under the conditions they faced. Early progress, however, was made with such achievements as the signing of the Canada–United States Great Lakes Water Quality Agreement in April 1972, and the securing of major funds for sewage-treatment loans under the Sewage Containment and Treatment Program approved by Cabinet on 18 October 1973.

But clear strides forward remained far more the exception than the rule. As one official noted, 'We always had major difficulties measuring the effectiveness of regulatory interventions.' The EPS remained an underresourced organization with not nearly the necessary information-gathering, pollution-measurement, and enforcement capacities. Also, the EPS was often involved in organizational conflict with the Inland Waters Directorate, the most logical source of water-pollution reconnaissance. Worst still, many industries, and the provincial governments who often came to support them, were not about to offer necessary cooperation and information that would allow the federal government to walk into cherished domains. As one senior official noted, 'We found considerable use by industry of the provincial governments. This was a very effective counter-attack as the provinces argued jobs, jobs, jobs! They were quite successful.'

Toxics and the Environmental Contaminants Act

Although there were some notable successes in the early period, the DOE was only two years old when a discussion process was launched for the development of new environmental regulatory legislation. By this time the EPS was becoming acutely aware that toxic chemicals were posing a serious threat to the environment and human health, and yet existing legislation was not well adapted to deal with this threat. A 1972 internal DOE discussion paper stated the problem as follows: 'media based legislation would not deal with the substance in all its physical states (i.e. solid, liquid or gas) ... Considering what we have been told about the pervasiveness of many environmental contaminants and their irretrievability in many cases, it is apparent that existing federal controls need expanding.'[11]

The Environmental Contaminants Act (ECA) was to be the DOE's response to the 'new' toxic chemicals coming into the market at an ever-increasing pace. Proclaimed on 2 December 1975, the ECA was to protect human health and the environment through the control of existing harmful substances and the prevention of new ones. Toxic contaminants become widely dispersed in the environment, last long periods of time in their harmful state, and are largely irretrievable once they escape and travel up the food chain in complex ways. Because of these characteristics and because biological changes in animals occur even in trace concentrations, it was wiser to try to control the release of them rather than to react after the fact. Furthermore, knowledge about the real human-health risks of toxic chemicals was and is extremely thin. The government stated that the act 'moves Canada further along the road from pollution clean up and role back toward the prevention phase of managing our use of the environment. The Act provides the power to acquire information about substances that may be harmful to human health and the environment and to take measures in consultation with provinces to prevent or control their use.'[12]

The road to the development and implementation of this type of substance-based legislation was a rough one to travel. The difficulties were revealed in a discussion document sent by Jack Davis to the provincial ministers of the environment on 18 April 1973.[13] In that document, Davis proposed legislation 'to prevent the widespread harm to the environment or danger to human health which will result if the release of certain substances (both existing and new) to the environment is not controlled.' But he also acknowledged the very real complexity of trying to employ this legislation within Canadian federalism when he stated, 'The power to implement controls under the Act is intended to be residuary.'[14]

The federal government was not going to try to move in areas where another level of authority was 'best able to do so.' Instead, Davis and the DOE wanted to 'require industry to furnish information on substances of concern in order that their hazards and benefits can be evaluated and recommendations made for controlling their release or escape to the environment if deemed necessary.' Otherwise, the DOE wanted to move only when 'there is no power to implement such controls under other federal or provincial legislation,' or when 'a province which has the power fails to act.'

Problems in administering the Environmental Contaminants Act were virtually inevitable. A 1979 background document starkly pointed out the magnitude of the task – namely, that 'some 300 new chemicals appear on the market each year. If each of these were screened for possible environmental and health effects, a judgement would have to be made on one chemical every 29 hours.'[15] But, as chapter 3 graphically showed, the EPS had experienced considerable reductions in resources, not the least of these in the areas of science and monitoring. The document continued by pointing out that 'to adequately screen one chemical for health effects alone by present techniques requires two years of testing and costs an average of $250,000 ... The Environmental Protection Service lacks the resources to be able to administer the *Environmental Contaminants Act* effectively.'[16]

However, when the ECA was developed, it was argued by Jack Davis that Canada needed the legislation because the U.S. Environmental Protection Agency was in the process of developing its own. Canada needed to guard against becoming a haven for polluting industries and the careless handling of toxic contaminants.

But, with the lack of EPS capacities for controlling toxic substances, Canada actually ended up relying heavily on the administration of toxic-chemicals legislation in the United States. Indeed, because of extensive U.S. ownership of the Canadian economy, Canada could free-ride, at least in part, on the United States for the testing, monitoring, and control of toxic chemicals.

Fish Wars: The 1977 Fisheries Act Amendments

The ECA and the Ocean Dumping Control Act, also passed in 1975, were developed just before the political-economic agenda shifted against environmental matters. The government's concerns over economic matters meant that federal pollution-control statutes were in many cases not simply implemented. For example, combined prosecutions under the main statutes never surpassed ten in total.[17] Enforcement under the Fisheries Act was also quite negligible, although there was more action in the Pacific Region than elsewhere.

The post-1975 period also saw the implementation of the 1974 Cabinet decision authorizing and encouraging federal-provincial accords in the area of environmental protection. Uncertainty about federal-provincial jurisdiction, a desire for intergovernmental cooperation, and increasing budgetary pressures, all pointed towards a retrenchment of DOE activities. The DOE, with the EPS taking the lead, developed the idea of a 'one window of delivery' approach whereby the federal government would establish 'base standards' and the provinces would see to their implementation. This, of course, meant that the EPS's enforcement capacities would be allowed to atrophy, and, as one former senior DOE official put it, 'The message got through. If EPS officials had tried rigorous enforcement, they would have got their fingers rapped.'

But even with the moves to become less of a 'regulatory burden,' establish 'one window of delivery,' and adjust to budget cuts, the DOE did launch a reform of the Fisheries Act in 1977. The reform initiative was, in large measure, intended to improve the habitat provisions of the act, but it also included pollution-control initiatives. The EPS, for its part, supported the Fisheries Act reform drive because it wanted a more flexible means to deal with new and changing pollution priorities. Whereas the 1970 amendments had allowed for regulatory control regimes rather than blanket prohibitions, the current act still, in the EPS's view, needed to incorporate more flexibility to address pollution on a site-specific basis. This was a further legislative elaboration of the negotiated form of regulation referred to earlier. Regulatory measures simply had to be developed around specific industrial processes and the market conditions of the firms involved. But there was also provision for requiring mandatory reporting and clean-up of pollution spills harmful to fish.

The response of industry and the provinces against the Fisheries Act reform initiative was swift and strong. But it was not so much the pollution provisions as those relating to habitat that attracted the ire of the provinces. The Fisheries Service sought new capacities to deal with development activities having a deleterious effect on the habitat or life-support systems of fish. The provinces saw this as a major infringement on their jurisdiction. One official added further that 'part of the reason for this increased conflict was also the expansion of the provincial environment ministries.' Competition among bureaucracies had become more intense.

It was in this period that Romeo LeBlanc lost any remaining enthusiasm he might have had for the Environment side of the DOE. His fights with the provinces were in many ways a distraction from his primary concern over the fishery. As one former senior DOE official put it, 'After the Fisheries Act amendments, LeBlanc did not want anything to do with [the] EPS.' The EPS was left in a situation where 'we did not do anything on regulation for almost a decade.' LeBlanc did not want to risk any more politically unattractive wars, and the Trudeau government, as a whole, had other major policy initiatives to pursue, as earlier chapters have shown. The federal withdrawal from environmental regulation was advanced to its most extreme point.

It took a series of environmental disasters to breathe life into the next environmental-policy initiative.[18] In late 1978, information about Love Canal in New York State began to surface. In November 1979, twenty-one railway cars carrying toxic chemicals derailed at Mississauga, Ontario, leading to the evacuation of 250,000 people. Also in 1979, there were media reports of a prolonged die-off of gulls, ducks, and shore-birds on the Toronto waterfront. Dioxins were found in Lake Ontario fish and in herring-gull eggs in several locations in the Great Lakes basin. And the level of toxic contamination of the Niagara River had reached crisis proportions as a result of groundwater run-off from haz-ardous-wastes sites. The public was soon to express increased concern about these and other incidents.

By the early 1980s, the DOE was also becoming more aware of the importance of building a policy agenda around high-profile public issues. Toxic contaminants emerged as a top priority in departmental planning exercises, and by as early as 24 January 1979, a new Toxic Chemicals Management Program was formed. Senior management were receiving an increasing number of public-opinion reports. One such report concluded that, 'as Canada's economic crisis extends into 1983, the public is making no compromises on the environment.'[19] By the first half of 1983, conclusions were that 'toxic chemicals and their suspected adverse affects on human health continue to dominate the major top-ics.'[20]

As a result of these rapidly changing conditions, the DOE began rethinking its position on federal-provincial regulatory responsibilities. A new legislative reform initiative for the Environmental Contaminants Act was also launched. John Fraser planted the seeds for these activities as early as 1979. He wanted the DOE to become more assertive regard-ing its environmental responsibilities to the Canadian public. Charles Caccia also actively pushed the DOE's legislative agenda in the year he had at his disposal. But it was Tom MacMillan who most decided to keep the idea of improved environmental-protection legislation alive in the Ottawa system.

The 'Toxic Blob' and the CEPA Process

Tom MacMillan became minister of the environment at a time when the process to reform the Environmental Contaminants Act was already two years old. Parallel federal initiatives on toxic contaminants had also been running for some time. For example, the Liberal government, in January 1984, had established a new Interdepartmental Committee on Toxic Chemicals to improve the coordination of federal activities in this field. DOE public consultations were proceeding, albeit slowly and quietly, motivated by increasing public concern over toxic contaminants. However, it was not long after MacMillan arrived in office that the entire range of federal government initiatives on toxics was transformed and broadened.

The death of the initial ECA reform drive followed closely on the heels of the August 1985 discovery of what came to be known as the 'toxic blob.' A mass of black oily substance was found floating near the bottom of the St Clair River, downstream from a Dow Chemical plant. It was later found that the 'blob' contained a veritable witch's brew of contaminants, threatening the drinking water of nearby Windsor, Detroit, and other communities. Later, the cause was identified as a leak from a waste-containment system at Dow's plant. Not surprisingly, the media coverage was extensive.

The Dow 'toxic blob,' however, had the impact it did because it followed soon after several other high-profile events which increased the public's awareness of toxic contaminants. Less than a year earlier, in December 1984, the Union Carbide pesticide-plant leak in Bhopal, India, had killed more than 2,000 people and injured 20,000 more. Early in 1985, a flat-bed truck carrying a large transformer from Quebec for storage in Alberta repeatedly splashed PCB contaminants over the road and surrounding vehicles while travelling through Northern Ontario near Kenora. Finally, in April 1985, the Inquiry on Federal Water Policy, appointed in January 1984 to examine the role of the federal government in water management, issued a preliminary report on its public hearings. The landmark final report of the inquiry, commonly known as the Pearse Report, was released in September 1985 and called for, among other important actions, better water-quality protection.[21]

These events made life very difficult for the new environment minister. MacMillan felt enormous pressure to do, and to be seen to be doing, something about an endless bombardment of environmental bad news. But, when he turned to his officials and asked what he could do about the 'toxic blob,' the answer was, as one senior manager put it, 'Nothing!' Environmental legislation did not empower him to act, and MacMillan 'was amazed that as environment minister he could not move at all!'

This frustration was critical in triggering what became a key DOE initiative during MacMillan's tenure. MacMillan stated his intention to get tough with polluters through a new generation of legislation. This startled many industry groups, and changed entirely the rules of the game between participants in the consultation processes. The stakes had been raised, and some three years later, in June 1988, the Canadian Environmental Protection Act (CEPA) was passed by Parliament.

The development of the CEPA, as previous experience had forewarned, was not a simple process. Many within the EPS were very unsure of whether the initiative should even be undertaken. As one senior manager noted, 'The development of legislation and the effort needed to push it through the process is very time-consuming, and usually nothing else can get done.' EPS officials were worried, given the political and economic conditions of the day, that an enormous expenditure in time and resources might be wasted on an undertaking with only a marginal chance for success. However, a massive growth in public concern and an improving economy soon convinced many within the EPS that the battle might be winnable.

The CEPA idea was given an substantial boost by the feedback that the Conservative government was getting from pollsters. By the winter of 1985, Decima, the Conservative pollster, informed the government that it was in serious trouble. It stressed that

public perception of the government has declined to the lowest point since its election. Dissatisfaction has increased *significantly* in the government's handling of inflation, federal-provincial relations and efforts to protect the environment. It could be concluded that there is the perception that the PC government is a "big business" government and places only secondary importance to social issues such as universality, taxation inequities, health and social services including the protection of the environment.[22]

The report went on to argue that 'environmental protection is not perceived as just another service ... but as a basic right ... Canadians value their natural heritage; they are greatly disturbed when any attempt to lessen existing environmental protection programs is contemplated or implemented; and it appears that questions concerning the environment are likely to become increasingly more sensitive and emotional at the political level due to their interrelationship with human health.'[23]

Public views had definitely hardened, and the federal government, along with business, was being seen as part of the problem and not as part of the solution. The federal government needed to show that it could get tough on polluters, and DOE environmental regulators were gradually being brought in from the cold.

In short, one of the key goals for the DOE in the CEPA initiative was to assert a more substantial federal presence in environmental regulation. The department was under intense pressure to rethink its 'one window of delivery' policy, which had been instituted after the Fisheries Act reform battles in the late 1970s. Experience had shown that, in many cases, neither the provinces nor the federal government had been willing to act on important pollution problems. The EPS was also left without adequate feedback from the provinces as to how well regulatory efforts at that level were going. The CEPA was going to be a key means for rebuilding EPS capacities, which had been allowed to wither badly, and Part II of the act provided information-gathering powers to force industry and the provinces to provide feedback regarding environmental regulations under federal jurisdiction. Table 6 provides a breakdown of the components of the CEPA.

The first draft of the CEPA was strengthened considerably through a major public-consultation process undertaken by the DOE. Departmental senior managers, who had become more sensitive to the use of public opinion as an ally, were urged by members of Parliament to provide for public input, and they made use of the opportunity. By December 1986, when the process was launched, the DOE had accumulated a pool of experience on consultations, such as with the Niagara process, discussed earlier in chapter 5. A key characteristic of the CEPA consultations was to provide funding for environmental groups. This allowed for a greater degree of balance in a process where industry and the provinces were very vocal and had special political-access points. On 25 November 1987, Tom MacMillan discussed his experiences in the public consultations with the Legislative Committee on CEPA Bill C-74. He stated:

Bismarck once quipped that 'laws are like sausages – it's better not to watch them being made.' ... I am inclined to agree with Bismarck. But the process is crucial, nevertheless, and I look forward to working with you in it ...

TABLE 6
Components of the Canadian Environmental Protection Act

PART I
Environmental-quality objectives, guidelines, and codes and practice (Regulations cannot be made under this part.)

PART II
Toxic-substance provisions, for:
- information gathering;
- controls on substances new to Canada;
- broad regulatory authority;
- interim orders for emergencies;
- clean-up of unauthorized releases;
- ministerial authority to direct remedial measures;
- export and import controls;
- control of lead in gasoline.

PART III
Nutrient control in cleaning agents (previously Part III of the Canada Water Act)

PART IV
Federal subjects' regulatory authority to control (pursuant to the provisions under the act):
- federal departments, boards, and crown corporations;
- federal lands;
- federal works or undertakings with federal legislative authority.

PART V
International air pollution (previously in the Clean Air Act)

PART VI
Ocean-dumping control (previously the Ocean Dumping Control Act)

PART VII
General provisions for:
- boards of review;
- agreements with the provinces;
- powers of inspectors;
- penalties and legal issues.

PART VIII
Consequential amendments and repeal of predecessor legislation

PART IX
Act proclaimed in force

Consultations led to substantial improvements in the bill. Cases in point are the inclusion of: the Ocean Dumping Control Act; greater opportunities for citizen involvement; increased public accountability of the Minister and of the government as a whole; the first ever definition of "environment" in federal legislation; a Priority Substances List; and biotechnology. The public called for all those changes and many others, and we responded. I am particularly grateful for the advice of environmental groups.[24]

The entire federal-provincial calculus, however, was then significantly altered by what one insider involved in the development of the CEPA called 'a last-minute agreement with the provinces to include the idea of equivalency agreements in the Act.' The provinces became very active at the political level, and with support from the Department of Justice, sought to control what they saw as a concerted federal charge into the provincial environmental regulatory field. Much of this lobbying occurred out of sight of the public-consultation process that had been ongoing and was nearing completion. One senior official noted, 'Quebec officials argued particularly hard for the equivalency agreements in defence of jurisdiction, and other provinces jumped on board.' Industry also encouraged greater checks on federal regulatory activities, and wanted to ensure that, where regulations were instituted, there was only one level of government to deal with and not two. The argument for achieving cost efficiencies through avoiding regulatory overlap became a powerful one to resist.

CEPA and Equivalency Agreements: The Rough Road to the New Millennium

The CEPA package was a considerable political achievement. But the key to its future operation is the equivalency agreements with the provinces. If an equivalency agreement exists, federal regulations under the CEPA are not to apply to that province. What constituted 'equivalent' standards was to be negotiated between EPS officials and the provincial regulators. According to the CEPA's Enforcement and Compliance Policy, agreements are to stipulate that every failure by industry or others to comply with equivalency regulations must be followed up by a provincial enforcement action. If the province failed to implement the equivalency regulations, the DOE reserved the right to develop and enforce federal regulatory standards under the CEPA.

However, the history of environmental regulation teaches us that these types of remedial measures are never easy in a highly political environment. Moreover, DOE officials were soon to discover how difficult it could be to negotiate equivalency agreements to begin with. Technical complexity and power politics can be used to extend the process of negotiation indefinitely. As one former senior official put it, '[The] CEPA may never work because of the success the provinces had in putting into place equivalency agreements. They became a Trojan horse. If you cannot reach agreements you cannot act.'

However, the DOE's CEPA agenda was far from totally usurped. EPS officials did achieve some important advances. One of the key motivations behind the CEPA was to establish better enforcement and compliance procedures. In the past, 'compliance agreements' had run into problems in the courts of law because, by design, they are inconsistent across industries, firms, and plants. This inconsistency was challenged on grounds of due process and equality before the law. The courts were not comfortable with an approach which, as noted above, reflected the difficult circumstances faced by EPS regulators. EPS officials sought to make headway in resolving this problem.

In the end, the CEPA was released with a companion Enforcement and Compliance Policy. The policy attempted to show how regulatory procedures would be more firmly grounded in the law. It also sought to address, in a highly visible way, public concerns about the legitimacy and effectiveness of the regulatory process from the federal government's perspective. Under the policy, and within the powers of the CEPA, regulatory inspectors are to be given the right to conduct inspections of firms and review materials, substances, records, books, and electronic data relevant to the administration of the act. If a violation is discovered, the inspector can choose one of a number of courses of action, depending on the severity of the case.

A 'letter of warning' or 'compliance direction' is issued for a minor violation when it is expected that the firm would undertake a prompt correction. If the offence is more severe, the inspector can request a 'ministerial order' or a 'court injunction,' or proceed to prosecution. A formal investigation, including search and seizure of pertinent evidence and information, could proceed when the case is deemed to be severe. Records of inspections, letters of warning, compliance directions, investigations, and other enforcement actions are to remain on the 'compliance record' of the firm. The consistent application of these procedures was intended to meet the court requirements for due process and equality before the law and to encourage industry to keep a good record.

However, EPS officials still had to deal with the legacy of the environmental regulatory process. Regulatory procedures, no matter how demanding on paper, could not be instituted without the political will to commit resources and to act. The advances made in the CEPA process did not make the complexity of the pollution problems, nor the old polluting industrial plants, disappear overnight. The DOE 'regulatory arm' had always been left woefully short of key resources and organizational capacities with which to wage the inevitable struggles for change. The resolution of these weaknesses remained very much an open question after June 1988.

Lingering issues included the fact that the EPS had always been largely a headquarters organization without the needed scientific and investigative personnel. When the CEPA was passed by Parliament, it included a Priority Substances List containing forty-four substances which must be investigated and scientifically tested for toxicity. Further pressure was added when the Green Plan committed the government to regulating up to forty-four substances over a five-year period.

Environmental groups and the Canadian Environmental Advisory Council (CEAC) had anticipated problems in this area and sought, as one official put it, 'a jump-starting of the regulatory process by coming forward with ten regulations concurrent with the passing of [the] CEPA.' Some groups even demanded a schedule for the regulation of listed priority substances. The CEAC option did not appear possible, given the powers at play within the legislative process, and the second option was opposed by the EPS because, as another official put it, 'it would have required continual revision of the act to update and revise schedules and substances lists.' A compromise was reached around specifying a Priority Substances List, which would be tackled as soon as possible. But, as previous experience indicated, the environmental regulatory process is very demanding on science and monitoring. Regulations can be seriously delayed if the regulator cannot bring solid technical capacities in these areas to the process, although these are not the only necessary ingredients.

Another lingering issue was the political, economic, and legal capacities of the EPS in the regulatory process. Owing to the historic shortage of these capacities, the EPS had always been at a considerable disadvantage in negotiations and political struggles with industry and other government departments and agencies. Environmental regulation includes a lot of complexity regarding the decision process and requires considerable political- and policy-process expertise to achieve success. The new CEPA also incorporated the then new federal requirement for a 'Regulatory Impact Assessment Study' (RIAS) to determine the socio-economic impacts of proposed regulations. These RIAS studies embody extensive economic complexity. Finally, multiple drafts of regulations must be produced for public consultations, and they must be defensible in legal terms. A massive shortfall in any of these organizational capacities could mean, as one ex-senior DOE official put it, 'another decade of stagnation.'

By 1993 an evaluation of the CEPA was under way, as provided by law. Few are pleased with the first five years of experience. Environmental groups bemoan the continuing lack of enforcement. Industry is lobbying against any implied or actual federal-provincial duplication. And the early annual reports on the CEPA scarcely provide any information that would allow dispassionate observers to assess as progress.[25]

Conclusions

In many ways, the long process of legislative catch-up that led to the passage of the CEPA is the most revealing of the DOE decision dilemmas precisely because its scope was so broad and compelling. Previously separate, single, media-based items of legislation (for water, air, etc.) had to be replaced and/or complemented by an overall multi-media and inter-media statute that sought to anticipate and regulate through the full cycle of pollution movements, from cradle to grave, from water, to air, to land, and back again in infinite circles and unpredictable currents.

Our account has shown that the DOE did catch up in some key respects and did improve its legislative capacity, but only to discover that catch-up was a game with no end and no real referees. The chapter has shown that three factors are vital to an understanding of the DOE and the EPS as environmental regulator and preventer. The first of these is changes in the scope and complexity of environmental problems. The EPS's regulatory agenda was altered over time as a direct result of the type and complexity of problems encountered, and how the general public were interpreting priorities. We can see, for example, how the impetus for the ECA, and later for the CEPA, was that existing legislation was not adequate for addressing the 'new' toxic contaminants. Toxics were becoming an increasing health threat and attracting broad press and public concern, particularly as related to Great Lakes water quality and the coastal fisheries.

The second factor is the difficulty in regulating fixed capital investment and powerful industrial interests. Much of the history of EPS environmental regulation can be explained by the character of the industrial processes and interests targeted for regulation. Companies with heavy fixed-capital investment in industrial processes cannot change overnight, often because of technological or financial limits. Time delay, with all of its environmental implications, is endemic in the environmental regulatory process. The exercise of political power by industrial interests, often aided by the provincial governments defending jurisdiction, has often exacerbated the time delay, particularly as related to job threats in single-industry towns.

Finally, we see the limited organizational capacities of the EPS as environmental regulator throughout its first twenty years. The EPS, as the DOE's 'regulatory arm,' has never possessed the resources adequate to the range of battles being waged. Resource scarcity, including the lack of money and staff for enforcement and knowledge gathering, has contributed mightily to a situation in which environmental regulation has never lived up to expectations.

11

Conclusions

The basic elements of the analysis of federal environmental policy have been completed. Our political-institutional assessment of green policy and decision making has been centred on the Department of the Environment (DOE) as an organization. In chapters 2 to 6, we have examined in detail the main political and institutional relationships that the Department of the Environment had to manage with environment ministers; other federal departments in the Ottawa system; provincial governments; business interests, environmental groups, and the public; and international institutions and other countries. We have also examined four major decisions, in chapters 7 to 10, which span the more-than-twenty-year period covered and which span key areas of the DOE mandate. In addition, we have examined several other decisions, ranging from the Green Plan of the early 1990s back to the Berger Commission of the 1970s.

What can be reasonably concluded about the evolution of federal environmental policy since 1970 and of its lead department? To provide this overall concluding assessment, we look again at the central questions posed at the beginning of the book. How has the federal environmental agenda been set over the past two decades? Where has environmental progress been greatest and where has failure has been most evident in the Canadian environmental record? How much of the success or failure of environmental policy can be attributed, not only to the DOE, but also to key political and institutional players, such as political parties, other government departments, the provinces, business and environmental groups, and international institutions? What capacity does the federal government have to advance the concept of sustainable development from its current status as a latent policy paradigm into a more entrenched one central to all future decision making? What are the keys to environmental progress likely to be in an age of increased international economic competitiveness and business power? And what are the prospects of communicating a green philosophy in an age of mass media politics and in relation to the potential for environmental issues to be a source of Canadian national unity?

In the final part of the chapter we revisit the overall framework used – namely, an approach which centres on what he have called the double dynamic of environmental policy. The first dynamic involves a search for political control among key institutions and the second involves a biophysical natural world where scientific uncertainty is increasingly present and intractable and where spatial realities are compelling. Concluding views are offered about how the conceptual basis of the book, building on other research and literature, improves the political-institutional basis for analysing environmental policy in Canada.

Federal Environmental-Agenda Setting

The analysis shows that the dynamics of setting an environmental agenda have been both episodic and multidimensional. First, there has been the 'rise and fall and rise' pattern, traced in the book. The political system took serious notice of environmental policy, first in the early 1970s and then in the late 1980s and early 1990s. In the long stretch in between, the political system was, purely and simply, interested in other things. And while conventional wisdom suggests that public opinion has always been ahead of politicians on environmental matters and therefore environmental issues ought to have been nearer the top of the agenda over the entire period, this is not necessarily so. The fact is that Canadians have, in reality, not been that far ahead of their politicians because their actual views on environmental matters are not revealed only through opinion polls. In other words, Canadians' environmental habits cannot, any more than can their politicians, be judged only on what they say. What matters is what they do. As consumers, members of interest groups and unions, voters for political parties, and polluters themselves, Canadians have in fact been closer in their environmental *actions* to their politicians than one would otherwise think.[1]

In part, the reason for the proximity in actions between Canadians and their politicians is that both learn about environmental matters in episodic rather than consistent ways. Thus a battle over acid rain or South Moresby is played out in an increasingly media-driven and media-screened political process that produces a decidedly uneven response. No one can argue that acid rain or any given new park is not an important environmental issue, but it is not clear that either was more important overall than, say, toxic chemicals as a whole, or the use of pesticides in agricultural production. 'High flyer' priorities such as acid rain reveal the double-edged nature of agenda setting. On the one hand, they help mobilize attention and support and, if handled well, can increase confidence in Canada's capacity to solve complex environmental problems. On the other hand, their stark political fame attracts resources and political attention away from other less-obvious or less-well-known hazards whose resolution may require a steady and continuous commitment of resources in a less spectacular way.

In any event, it is evident that what gets put on the environmental agenda is clearly not produced by a tight causal link between public opinion and politicians. The power of special interests is also pivotal, whether it be the chemical or pulp-and-paper industries putting pressure on Ottawa to lay off the heavy hand of regulation, or ENGOs strategically targeting a conveniently occurring environmental-assessment hearing to advance their agendas.

Environmental Successes and Failures

While green agendas are the ultimate outcome of a pluralist struggle for resources over a wide range of issues, the environmental agenda from 1970 to the early 1990s has not been entirely unplanned and random, and there have been important successes. The role of science and scientists and the impact of various efforts to plan an agenda have been important in this regard. Thus the Great Lakes Agreement emanated from an early scientific consensus within the DOE, and indeed, before the DOE's establishment, that Great Lakes pollution required priority attention. As a result, progress has been made in cleaning up the Great Lakes, although new problems are emerging.[2] Early general progress on air pollution was also attributable to extensive scientific and federal-provincial cooperation.

Progress has also been aided by what we referred to in chapter 6 as international paradigm-setting activities. The original impetus of the Stockholm Conference of 1972, the 1980 World Conservation Strategy, and the articulation of a 'sustainable development' paradigm by the Brundtland Commission in the mid-1980s helped nudge the green agenda forward to its early 1990s peak at the Rio Earth Summit. These were in turn related to, and aided by, the DOE's own internal planning exercises examined in chapter 3.

Even the opportunistic 'pouncing' activities of ministers examined in chapter 8, on Parks Canada, produced genuine environmental advances, not to mention a certain micro–priority setting rationality of its own. Ministers found parks a congenial part of the mandate and could use the creation of parks with a certain amount of symbolic flair. Once established, parks were almost always genuine environmental advances. Moreover, they were projects that had a definite beginning and end. And many ministers were attracted to this sense of concrete achievement, especially because so many other aspects of the DOE mandate were swampy and indeterminate and one could scarcely tell from one year to the next whether things were getting better or worse.

In still other areas, such as toxic substances or forestry, or in the need to price water resources to reflect environmental costs, advocates from within and outside the DOE saw little for their efforts. Concern would appear in planning documents but not in actual resources and power.

Hence, there exists to this day a vast range of priorities where past inactions leave a growing unfinished agenda. Meanwhile, new issues are discovered, such as global warming whose scope and depth of difficulty vastly exceeds those of its predecessors. Given this history, and its years in the political doldrums, it is little wonder that the DOE seized the moment of opportunity when it emerged in the late 1980s and launched its 1990 Green Plan. Given its chequered past, it saw no alternative but to reach for and grab as much as it possibly could when the political stars looked favourably upon it in 1989–90.

In this context, the reaction to the Green Plan and to the processes that led to it were absolutely predictable and quite understandable.[3] The Green Plan contains many priorities that make sense but that are simultaneously impossible to carry out fully in a few years. The ENGOs were angered by the DOE's early secretiveness in developing the Green Plan and pressured it into holding a public-consultation process. Most ENGOs regarded the Green Plan's content as being insufficiently radical. But, at the same time, the ENGOs knew that the DOE was being attacked mercilessly from within by ministers, departments, and interests which did not want to see the DOE garner the large amounts of money and decision-making powers contained within the Green Plan.

Political Parties, Key Institutions, and Future Decision Making

As discussed further below, it is not difficult to conclude that the DOE's capacity to manage and control its main relationships with other institutions has been, on balance, weak. In many respects this weakness resulted from the DOE's own internal conflicts over ideas as to where it wanted to go and to a lack of consistent leadership. But in an even larger sense, the failure over the period as a whole to make environmental issues a central part of governance in Canada must be attributed as well to the other institutions examined.

Chief among those responsible for the slow greening of Canada are Canada's two federal political parties and the prime ministers who headed them. The Liberal record began hopefully and with considerable international leadership. Both Prime Minister Trudeau's initial personal interest and Jack Davis's leadership paid some early dividends. But, thereafter, the Liberal interest faded badly. For their part, the Mulroney Conservatives began with the disasterous Blais-Grenier period and improved under the tenure of Tom MacMillan, but the DOE only caught significant prime-ministerial attention in the late 1980s when Brian Mulroney appointed his then close ally, Lucien Bouchard, as minister of the environment.

In both political parties' long lag periods, they hitched their environmental political star primarily to the acid-rain issue, in part for actual environmental reasons but often so that they could get at least some symbolic foreign-policy leverage or media credit over the Americans. While both parties spent money on acid rain in particular, they also cut resources for the DOE in general, and shuttled mainly junior ministers in and out of the portfolio in a series of acts of benign neglect. This does not mean that there were not environmental stalwarts among the ministers. Jack Davis, Charles Caccia, John Fraser, Tom MacMillan, and Lucien Bouchard made a difference, had some kind of a personal green agenda, or grew in the job. But the spurts of leadership were too few and far between.

As a result, other institutions, already ill-disposed to cooperate, knew that the DOE was vulnerable and acted accordingly. In a pluralist political system, these institutions also had their own bases of power and their own larger statutory and political mandates and causes. A key issue in their relationships with the DOE was over how decisions were made and, as well, over their content and impact on special interests. Most environmental decisions over the period from 1970 to 1990 were curative, that is, intended to clean up or mitigate effects after the fact. As a result, green-decision processes sought to challenge decisions already made, an invitation to bang heads against concrete walls if there ever was one.

Only in one or two areas was anticipatory and preventative green decision making tried. In the regulatory process under the DOE's Environmental Protection Service (EPS), regulations applied to new plants were partially anticipatory and therefore encountered somewhat less resistance than did those for old plants. In addition, there was the regime for the environmental assessment of projects. In principle, such assessments are anticipatory and seek to influence choices at the front end of the decision cycle. This too, as we have seen, was resisted. But for environmental assessments, there was also a learning curve in evidence. Some projects were cancelled, altered, or rescheduled. And by the early 1990s it was ENGO-initiated court cases that were forcing the government to legally entrench the assessment process.

Future relationships among the key institutions will be dependent upon what decision-making processes are put in place in the mid-1990s and beyond. Indeed, to institutionalize many aspects of a sustainable-development philosophy, 'process' is close to being 'everything.'[4] Accordingly, reforms such as the non-statutory requirement (via a Cabinet order) for all policies, not just projects, to be given prior environmental assessments will warrant close analysis.

While prior environmental assessments of policy are clearly necessary if there is to be any hope of entrenching environmental values in decision making, undertaking them is by no means a sure or easy thing. Quite apart from the bureaucratic trench warfare that is bound to occur, there are genuine problems regarding the acquisition of adequate expertise and operational data on environmental effects. There are also real philosophical and scientific divisions over just what sustainable development means operationally in any specific policy circumstance.

Sustainable Development as a Latent Policy Paradigm

The concept of sustainable development can be considered in the 1990s to be a latent policy paradigm. A well-developed policy paradigm provides a series of principles or assumptions which guide action and which suggest solutions in a given policy field. It becomes entrenched in the thought processes and education of various professionals and interests in a given policy network. Despite its decade-long political existence, sustainable development is a latent paradigm primarily because it does not yet have the coherence of earlier paradigms such as Keynesian economics or the liberal economic-growth paradigm.[5] But a workable paradigm for new green-decision processes is essential if day-to-day Cabinet decision processes, not to mention the decisions of private industry, are to be altered. But clearly not just any glitzy idea will do if environmental departments such as the DOE, working with and through other players in their policy sector, are to move from being a centre for micro-governance to one of macro-governance.[6]

Sustainable development has considerable staying power as a paradigm in the 1990s and beyond for several reasons. First, though it has ample constructive ambiguity in its meaning, it has a political and intellectual resonance that is quite genuine. First, it reflects the cumulative pressure and debate fostered by a global political constituency that spans the Green Parties, traditional conservation movements, and ecological groups. Second, it reflects a more thoughtful reformulation of an earlier zero-growth paradigm in the 1970s that deservedly went nowhere politically. Third, it acquired an initial international legitimacy among scientists and resource-management experts during its gestation and progress, as we saw in chapter 6, through several international forums. These ranged from the International Union for Conservation of Nature and Natural Resources (IUCN) to the OECD, and from the Brundtland Commission to the Rio Earth Summit. And last, but hardly least, it undoubtedly strikes a chord among ordinary people who appreciate that the world environment is deteriorating and who express concern about the kind of planet their children have to live on. Obviously this latter mass resonance does not reflect any detailed understanding about what sustainable development means. Rather it reflects an understanding that something is wrong and that future approaches cannot be the same as those adopted in the past.

While recognizing some of its vagueness, those who defend the genuineness of the new paradigm in the longer history of ideas point to several features which make it different from prevailing economic and environmental orthodoxy. The first feature is the explicit 'entrenchment of environmental considerations in economic policy making.'[7] This was not acknowledged in the past the way it now is, even at a conceptual level. The second feature is that sustainable development evokes the language and concerns of the Third World and implies a consideration of needs and equity between North and South and also intergenerationally within the North and South. The third feature is that development does not equate with simple economic growth in that it acknowledges the non-financial aspects of economic welfare.

Thus when G-7 leaders endorsed the concept of sustainable development at the 1988 Toronto Summit they did so because the paradigm had a minimal but growing legitimacy. Its acceptance as a respectable paradigm has also been aided by the fact that, in contrast to the self-interest ethos of the 1980s, it seemed to be reaching out to the common problems of mankind.

The concept of sustainable development thus has an important foot in the door of central governance in Canada and elsewhere. Ideas are thus an important starting-point, but they are not a sufficient condition for future environmental success. Sustainable-development ideas must also be workable in a quite detailed analytical and institutional sense. Within the federal governent and its departments, hundreds of policy decisions and thousands of project decisions are made. Environmental analysis, including its vital scientific content, must feed into Cabinet committees and decision memoranda in a way that is both timely and coherent. There it will join values, pressures, and interests also being brought to bear on increasingly beleagured ministers. Sustainable development as a basis for assessing all federal policies will remain a latent paradigm and not an overarching one unless it assumes a more concerted operational relevance. This is not yet happening in the Ottawa system.[8]

Greening, Economic Competitiveness, and Business Power

The first twenty years of federal environmental policy is replete with instances where the DOE had to accommodate itself to the realities of economic and business power. From Jack Davis's bridge on the Burrard Inlet, to acid rain, to pulp and paper, and in myriad environmental assessments, the environment was viewed as a trade-off with economic development, and the latter won far more often than it lost. However, during the latter part of the 1980s, the relationships between environment and economy matured if not to a full reconciliation, then at least to the beginnings of a better mutual understanding.

The nature of this improved understanding grew out of many factors, including the unrelenting pressure of the ENGOs, globally and nationally; the mutual learning and semblance of somewhat greater trust that built up between ENGOs and business in the various DOE stakeholder consultative processes; the articulation and initial endorsement of the concept of 'sustainable development' as an alternative to the 1970s no-growth ethos of many ENGOs; and the changing nature of environmental coalitions within the business community.

These changes within business, as we saw in chapter 5 and elsewhere, reflected a newly evolving four-part breakdown of business interests regarding the environment. Business is no longer a monolithic anti-environment force. By the early 1990s the business community contained old-style polluters; industrial sectors that were seeking to take advantage of their clean image; new environmental industries that stood to gain from the sale of environmental technologies and green consumer products; and financial circles increasingly vigilant about the financial and insurance consequences of the environmental risks of their client companies.

These changes in the structure of business interests were accompanied by an equal change in the degree of complexity among the ENGOs and among the global coalitions that were forming. It was not only business and finance that went unprecedentedly global in the 1980s and early 1990s. So too eventually did the environmental movement. Indeed, a good case can be made that the Green Plan was propelled more by a need to keep up with international pressures than a need to assuage domestic ones, however compelling the latter had become.

All of these developments are basically grounds for optimism regarding environmental policy and competitiveness in the 1990s, but they still leave untended in the Canadian context several overall imperatives regarding greening and the economy that the DOE and all Canadians must face. To see these, it is useful to return to the Green Plan, both for what it contains and reveals about Canada's environmental history and for what it fails to grapple with or at least discuss with sufficient candour. The first imperative deals frontally with the competitivess issue and economic-efficiency concerns and involves the use of market instruments as a complement to traditional regulation.

One of the accurate charges levelled against the DOE is that it has been far too often an economically illiterate department. It has failed over most of its existence to consider seriously the role of market instruments such as taxes, charges, and so-called tradeable-pollution permits to augment and complement its traditional regulatory urges.[9] The main reason that such more diverse instruments of policy are increasingly needed is that there are simply too many permutations and combinations of environmental situations for any one player to 'command' results. If market approaches are used, they can become for private interests and citizens the functional equivalent of what the prior environmental assessment of Cabinet policies is for government. In short, they are a vital device for 'costing the earth' and for inducing and requiring numerous interests to make their decisions with a greater practical realization of actual costs to the ecosystem and to the economy but with a high degree of flexibility as to how various interests take appropriate action.[10] Without such instruments, there is little chance of environmental progress being made. Prior to the Green Plan, the DOE hinted at its recognition that this was increasingly important but then gave it only the barest of mentions in the Green Plan itself. Only in 1993 were modest beginnings made in the form of DOE-funded research on economic instruments.

Communicating a Green Public Philosophy and Canadian Unity

A final issue of importance regarding both the past and the future greening of Canada concerns whether green plans or related green public philosophies can be politically sold and communicated in a mass-media political age and whether they can be an important base for strengthened national unity in Canada. More precisely, the question emerges as to whether the Green Plan or a sustainable-development paradigm should be considered for the century's last decade, the functional equivalent of the postwar Keynesian revolution, albeit perhaps this time without a Keynes. There are grounds for partially viewing the Green Plan in this way in that, at its core, as we have seen above, it is an attempt to shift the whole paradigm of all public-policy debate.

In broad terms, the Green Plan does call for a sea-change in values and in the way all institutions make decisions. But unlike the Keynesian revolution and the accompanying social agenda of the postwar reconstruction period, there is now no Keynes whose message can be easily communicated. By this we mean that, in this era of cynical mass-media-driven politics, there is no fairly easily communicated and understood way in which sustainable development or green plans can be politically communicated and put into operation. To some extent in the earlier period, Keynes had supplied not only a clearer paradigm for macro-economic policy but also a basis around which quite diverse interests could rally in a somewhat simpler era of political communication.

Moreover, because there is no simple cause-and-effect pattern in the environmental-policy equation – the double dynamic ensures that there is not – the Green Plan must resort to all-embracing concepts such as as 'partnerships' and 'planning for life.' Moreover, it is not possible to attach easy or conventional 'left–right' political labels to greening. If the Keynesian period was mildly left of centre in its political orientation and could help invoke the evils of capitalism as it was practised in the 1920s and 1930s, this is less feasible for the political economy of greening. Capitalism still has its environmental villains to be sure, but so does the state: the citizen as voter and consumer and the worker as producer. The revolution in the former Soviet Union and in Eastern Europe has shown graphically that pollution is at its worst where there are no clear property rights.

All of the above is made even more difficult by the rampant cynicism that invariably seems to greet any governmental initiative. Modern mass-media-based politics, moreover, has little tolerance for problems that take years to be defined and solved.

These realities are also important in terms of whether the greening of Canada can become the central basis for a 'national project' with beneficial implications regarding the evident disunity of Canada in the early 1990s. In Richard Simeon's phrase, a national project can be thought of as a major, morally uplifting or unifying initiative that gives Canadians from all regions and classes a sense of pan-Canadian achievement.[11] Examples of such projects and achievements in the past include the construction of the immediate postwar social-welfare state and the achievement in the 1960s of medicare and the Canada Pension Plan. It is thought by many that the early 1990s constitutional battles, which came on the heels of a highly divisive free-trade battle, have caused Canadians to lose a collective sense of themselves. They have lost a sense of national unity and of achieving national projects together. This is also linked to growing views that the national government has been weakened in favour of markets and of the provinces.

While the greening of Canada has considerable potential to be such a national project, and indeed there may well be a deep yearning by Canadians to be more at one with nature and its limits, there are also serious obstacles. The environmental issue, as a series of tens of thousands of practical choices, may not be nation defining in any conventional sense. It could certainly not be modelled on any simple notion of assertive federal government authority. The areas of needed action, of the melding of area and power, are dispersed nationally, locally, and internationally.

As the history of federal environmental policy testifies, there is no such thing as green political bliss. Environmental issues and sustainable development will loom ever larger in the late 1990s and beyond, but they are no guarantor of national unity. But they do require sustained action and attention by all institutions in Canada and in a shrinking and interdependent world.

The Double-Dynamic Framework and Environmental-Policy Analysis

A related second set of concluding observations concerns the overall conceptual case made in this book for the basic framework employed. We have argued in both the structure and the content of this book that there is a compelling double dynamic at work in federal environmental-policy making. The overall story of Canadian environmental policy and of the DOE's pivotal role in it involves a growing tension between two dynamics: first, a time-consuming search, and indeed often a long series of power struggles among key institutions to agree on what should actually be done about environmental problems and policy; and second, the ever-changing biophysical realm which is characterized increasingly by unpredictability, scientific uncertainty, and stark spatial realities as the environmental problems of humankind, plants, animals, land, air, and water blend together with great speed and growing degrees of complexity. As the June 1992 Rio Earth Summit showed graphically, both dynamics are also now unambiguously global as well as national, regional, and local.

The DOE and the Power Gap

The DOE's struggle with the first dynamic is not an abstract notion. To determine the nature of the DOE's influence and capacity, we have had to look in the first six chapters at the key political-institutional relationships involved. More particularly, we have examined the presence or absence of all of the concrete organizational resources needed within a policy sector for it to be effective politically and substantively: a coherent set of workable policy ideas; political and bureaucratic leadership; money and financial resources; legal jurisdiction and statutory capacity; and scientific and technical knowledge. As we have stressed throughout, it is the possession of all of these attributes, not merely some of them, that is necessary to consolidate and advance environmental goals and to understand properly Canada's environmental-policy successes and failures.

While the concept of a policy sector within the state has some links with the concept of policy communities, it inherently begins more within the state, and, as we have suggested, analytically reaches outward. It requires us to look more concretely at the above-mentioned organizational resources within the state and hence at key bases of actual political power and the power to act effectively in substantive environmental terms.

In the realm of the first dynamic, the DOE sought to establish its power over the other institutions involved, but for the most part could not. In the realm of ministerial relations, and hence political and bureau-cratic leadership, the DOE suffered from a high turnover of ministers and from organizational leadership that could never sustain itself for very long. The DOE was for too long a house divided, with some units actively campaigning to leave and others reluctant to join. The fact that there were deep philosophical divisions and hence unworkable and non-unifying policy ideas between those who advocated integrated resource-management approaches and those who sought direct regula-tory solutions was both clear and understandable. Indeed, these com-bined approaches lodged in one agency is partly what made the DOE different from its U.S. counterpart, the Environmental Protection Agency. The EPA was more clearly a regulatory than a resource-manage-ment agency. But the dual philosophy for the DOE was also ultimately debilitating since neither approach, given the DOE's other constraints, was even half-practised.

In the interdepartmental realms of the Ottawa system and in federal-provincial relations, the DOE suffered from equally severe weaknesses, though clearly not always of its own making. The political centre in Ottawa, and other departments and ministers (and the interests they represented) were basically unwilling to give environmental matters a high priority. This was reflected in, and exacerbated by, the DOE's declining resource situation, and in the weakening of its scientific and technical capacity during most of the 1980s. Vis-à-vis the provinces and with regard to handling inter-media (air, water, and land-based) pollut-ants, the DOE never actually possessed a workable legal jurisdiction or statutory capacity. When its money and scientific capacity were also allowed to atrophy, it lost still further bases of federal-provincial and intra-Cabinet influence. The CEPA legislation was certainly an improve-ment on an excessive reliance on the Fisheries Act, but in the early 1990s the actual efficacy of the CEPA is still unproven, and hence the DOE's statutory capacity is still questionable.

These core policy-sector organizational resources are central to the approach to environmental-policy analysis that we have adopted. But elements of the policy-community approach are also vital to any overall explanation. Thus with regard to its core political clientele – business, the ENGOs, and the public – the DOE was never able to work out a con-sistent *modus vivendi*. With business, it was far too cosy in its early regu-latory processes but at the same time was never willing to acknowledge the business sector's criticism that the department was too often eco-nomically illiterate. With the ENGOs, it was not until the mid-1980s that

the DOE seemed ready consistently to see the ENGOs as a potential natural ally. Until then, the DOE was too intimidated by the range of ENGO
opinion, by the fear of angering business opinion if it cozied up too
much to the ENGOs, and by the fear that it could never fulfil the ever-
expanding expectations and agenda of the ENGOs. As far as the public
was concerned, the DOE knew that it always had a broadly supportive
general public but could only rarely use such support to mobilize either
its own or other ministers.

As for the ENGOs themselves, there is no doubt that they were the
main advocates of change throughout the period. Without them, environmental progress would have been even more limited. The ENGOs,
however, also suffered from a cautious relationship with the DOE, never
quite wanting to support the department fully or too openly, even during the Green Plan process, and even though it knew that the DOE was
its best hope inside the government. As chapter 5 showed, the ENGOs
also laboured under the yoke of inadequate financial resources but also
under considerable philosophical division between moderate reformist
nature groups and groups with quite radical perspectives, both as to tactics and as to desired end results.

In the international realm, the DOE seemed more at home, especially
in conveying environmental idealism and leadership in the multilateral
arena. This was broadly true throughout the period, from Stockholm to
Brundtland, to Rio. But, as international hazards and issues became
more complex, the DOE's international relations became caught in a
more intricate web of domestic political interests. In general then, its
international role is one of the DOE's main points of success. However,
in the bilateral Canada–United States arena, after a good start with the
Great Lakes Agreement, the realities of power politics with the United
States certainly took hold. Here the DOE undoubtedly deserves some
sympathy because the Reagan Administration played unremitting political hardball for all of the 1980s, leaving progress on acid rain to await
the arrival of the Bush Administration and a Democratic party–controlled Congress. Under the Clinton Administration, Canada may well
seem like the more hesitant continental environmental partner, though
this remains to be seen.

The dynamics and some of the content of the Green Plan and the Rio
initiatives suggest that the DOE has come closer in the early 1990s to getting its policy act together. But the verdict for the first two decades of the
DOE's capacity to deal with the first dynamic is quite clear: it was weak
at worst and grossly uneven at best.

Biophysical Realities and Scientific Uncertainty

The second dynamic refers to the even more difficult challenge of coming to grips with biophysical realities, scientific uncertainty, and daunting spatial realities. This dynamic clearly affects the specifics of defining and dealing with key environmental problems. In short, it deals with the task of actually influencing nature, either to help restore it to a renewed sustainable state or to harness its own recuperative powers. The second dynamic inherently poses questions such as: How can environmental agencies systematically track the flow and impact of toxic substances as they move up the food chain? How is the biodiversity of nature's plant and animal life accounted for and preserved on a global basis? It is in this kind of biophysical realm that scientific and technical capacities are vital if one is to understand and respect nature and if one is to appreciate the limits of mankind's capacity to act.[12]

We argued in the introduction to this book that all policy fields face some kind of physical dynamic of this kind, especially departments dealing with natural resources, such as Agriculture, Fisheries, Forestry, and Energy.[13] But despite this partial parallel in other policy fields, it remains true that environmental policy is unique in the scope and degree to which this second dynamic goes to the heart of its very existence. Environmental policy at its core has always had radical implications for mankind precisely because it ultimately does seek a reconciliation with nature across the board and at the same time challenges the assumptions of all existing institutions.

We have also attempted to make the second dynamic as explicit a part of our framework as possible because our reading of existing Canadian and comparative literature on environmental policy simply fails to deal with it adequately. Other environmental-policy scholars are certainly aware of its importance, but too often for our liking treat it too much as mere context.

A more complete inclusion of the second dynamic also helps show why the development of scorecards for environmental progress is characterized by such extremes of opinion. It helps indicate why reach always seems to exceed grasp and why any reasonable judgment of federal environmental policy must honestly characterize the past as the begrudgingly slow and gradual greening of Canada. Even though we can usefully speak of the DOE having had an initial rise in the early 1970s, a long fall for the next decade or more, and a renewed burst symbolized by the 1990 Green Plan and the Rio Summit, the picture over the whole period since 1971 is not a reassuring one. The reasons, as we have seen, are not hard to find, and they embrace key features of the second dynamic.

The second dynamic starts with the essential facts about ecosystems – namely, that they are in part self-regulating communities of energy and matter, that they perform essential services for economies and for human and other life-forms, and that they are governed by the laws of thermodynamics.[14] These laws include the fact that matter and energy cannot be destroyed or created and that economic production merely transforms them from one into the other as wastes. These laws also suggest that when such waste transformation occurs, the resulting waste products, including valued consumer products, have a lower entropic value – in short, they are more disordered and less useful than in their previous natural state.

The second dynamic also includes the presence of scientific and technological uncertainty and controversy both in addressing substantive green issues and in mobilizing the role of science in public-policy formation and debate about them.[15] Basic issues of cause and effect, risk assessment, hypothetical occurrences, and the limits of science are more prevalent and endemic in the environmental field than in other policy fields. The case-studies of major decisions in chapters 7 to 10, as well as other smaller case-study decisions in chapters 2 to 6, reveal the second dynamic in several ways.

The analysis of the acid-rain decisions showed that, although there is a tendency with hindsight to think of acid rain as a simple hazard, it in fact was and is complex. First, the core of the hazard, SO_2, is but one of several interacting pollutants involved in the long-range transport of airborne pollutants. The chapter on acid rain showed that it took several years to generate a sufficient minimum consensus among scientists that there was a problem and several more years for proper baseline data to be established. Uncertainty was enhanced by the fact that there were different sources of pollution in the United States and Canada, and that preferred control technologies differed.

Similarly, in the chapter on the CEPA, and in other references to industries such as pulp and paper and chemicals, it can be seen that solutions are limited by at least two biophysical, technical, and spatial realities. The first is that there really are differences between old and new production plants and between the configurations of water bodies and production circumstances in hundreds of different sites across a vast continent and encompassing dozens of cities, not to mention one-industry towns and communities. The second is that, even in the 1990s, science is still limited in its capacity to assess the assimilative capacities of different bodies of receiving waters. There remains as well scientific uncertainty regarding particular toxic effluents from hundreds of manufacturing processes, including a large number of unidentified substances. Our account of environmental regulation in chapters 1, 3 and 4 also showed that scientific uncertainty is inherent in the whole debate in environmental regulation about the 'end of pipe,' best-practicable-technology approach. Such an approach is based on an admission that the alternative approach, assessing the assimilative or loading capacities of each body of water and setting standards accordingly, is scientifically too complicated in many situations. And yet these difficult or impossible problems pale when compared with the problems of understanding even larger fish/forestry habitat issues.

If these realities are true for acid rain and regulating effluents, then they are even more the case for more complex hazards such as NOx–VOCs or global warming. While the latter were not accorded chapter case-study status, they were discussed enough in the book to indicate that the problems are even greater, not only regarding the interdependence of multiple hazards but also with respect to the complex array of interests and countries affected.

The other case-study chapters show the variety of circumstances involved in coping with the second dynamic. The South Moresby case-study was by no means atypical in Parks Canada's agenda. The dilemmas of ensuring that ecologically vulnerable or representative areas are preserved in particular Parks, and in the parks system as a whole, is a momentous task. The inherent physical expanse of the parks is daunting enough. But the South Moresby case showed the complex linkages between logging and fish habitat and the preservation of species essential to Haida life and traditions. The chapter revealed Parks Canada's desire to 'protect through science' and to enhance the role of parks as environmental benchmarks for gene pools but also concluded that such work has only barely begun.

The two case-studies in chapters 9 and 10, involving decisions about the search for viable legislative frameworks, one on environmental assessments and the other on regulatory protection through the CEPA, reveal the second dynamic in somewhat different ways. Both show the sheer problem of horizontal and geographic-spatial scope that these basic kinds of environmental policies require. Whether in the anticipatory and preventative mode, in the case of environmental assessment, or in the clean-up and regulatory aspects, in the case of the CEPA, environmental-policy makers face the need to somehow intersect with and change behaviour at literally tens of thousands of project and decision-making sites. There is a daunting sheer physical immensity to these essential tasks of social or horizontal regulation and management through legal and incentive-based policy frameworks and sequences of action.

Environmental assessments also require truly enormous amounts of scientific and technical analysis. They are also still fundamentally threatening to the status quo. Assessments were resisted for so long, in part because they were administratively complicated, but more fundamentally because they truly did have radical implications for entrenched decision-making interests. Similarly, the CEPA, though it is a necessary step in strengthening federal jurisdiction, must confront severe physical limits. The negotiation of equivalency agreements with the provinces is difficult not just because bureauracies are sluggish but because the actual content of equivalency agreements requires detailed kinds of knowledge about technical standards which in many instances simply do not yet exist.

The further point to stress is that the tension between the two dynamics is growing, not abating. That is, even if there is improvement in closing the political power gap, there is a growing tension because the scale of ecosystem interdependence and scientific uncertainty implies, as in a football analogy, that the goal posts are not only always being moved but are also being moved farther away.

Thus, in terms of the overall test inherent in coming to grips with the double environmental dynamic, or even in establishing itself as the political engine of a policy sector, the DOE cannot be considered a success. But is this a fair test? After all, we seem to be saying that if the DOE cannot consistently control other institutions and cannot tame nature in twenty-plus years, then it is a failure. In one sense, this is a fair test because it deals with the ultimate radical agenda that environmental issues represent. Environmental policy is, at its core, no mere regular, garden-variety policy field. Even the most conservative environmental advocates seem to want major change. Therefore, relative to this ambitious agenda, has much change been achieved? The answer must be no.

In other respects, however, the test inherent in the double dynamic is unfair in that it fails to introduce a 'degree of difficulty' criterion. Moreover, if it is a test that few can meet (other departments, other countries) or that may be impossible to meet, then it does not do justice to the areas where progress has been made or to dedicated DOE, ENGO, and other business and governmental practitioners who helped make some good things happen. Staunch defenders of the DOE or of Canada's environmental record might well say, 'Judge us against the situation in 1971 or in comparison with other countries, not against a test made in heaven.'

There is little doubt, however, that for both Canada and the planet, both halves of the double dynamic have to be understood and addressed with a sense of sober hope, sustained thought and analysis, and continuing political commitment and action by all social institutions.

Notes

INTRODUCTION

1 World Commission on Development and Environment, *Our Common Future* (Oxford: Oxford University Press 1987), 43.
2 See G. Bruce Doern and Glen W. Toner, *The Politics of Energy* (Toronto: Methuen 1984).
3 The discussion in this section is derived from Michael Jacobs, *The Green Economy* (London: Pluto Press 1991), ch. 1.
4 See David W. Pearce and R.K. Turner, *Economics of Natural Resources and the Environment* (London: Harvester Wheatsheaf 1990).
5 See Jacobs, *The Green Economy,* ch. 1.
6 See G. Bruce Doern, *Science and Politics in Canada* (Montreal and Kingston: McGill-Queen's University Press 1972) and G. Bruce Doern, *The Peripheral Nature of Scientific and Technological Controversy in Federal Policy Formation* (Ottawa: Science Council of Canada 1981).
7 Robert Boardman, ed., *Canadian Environmental Policy: Ecosystems, Politics and Process* (Toronto: Oxford University Press 1992).
8 See Robert Paehlke and Douglas Torgerson, eds., *Managing Leviathan: Environmental Politics and the Administrative State* (Peterborough: Broadview Press 1990), Robert Paehlke, *Environmentalism and the Future of Progressive Politics* (New Haven: Yale University Press 1989), Doug Macdonald, *The Politics of Pollution* (Toronto: McClelland and Stewart 1991); Michael Whittington, 'The Department of the Environment,' in G. Bruce Doern, ed., *Spending Tax Dollars* (Ottawa: School of Public Administration, Carelton University, 1980), ch. 7; and O.P. Dwividi, *Resources and the Environment: Policy Perspectives for Canada* (Toronto: McClelland and Stewart 1980).
9 See Ted Schrecker, *The Political Economy of Environmental Hazards* (Ottawa:

Law Reform Commission of Canada 1984), Ted Schrecker, 'Resisting Regulation: Environmental Policy and Corporate Power,' *Alternatives*, 14/1 (1987), 9–21; and Kernaghan Webb, *Pollution Control in Canada: The Regulatory Approach in the 1980s* (Ottawa: Law Reform Commission of Canada 1986).

10 See George Hoberg, 'Sleeping with an Elephant: The American Influence on Canadian Environmental Regulation,' *Journal of Public Policy* 11 (1991), 107–31, and George Hoberg, 'Environmental Policy: Alternative Styles,' in Michael Atkinson, ed., *Governing Canada: Institutions and Policy* (Toronto: Harcourt Brace Jovanovich 1993), 307–42.

11 See Glen Toner, 'The Canadian Environmental Movement: A Conceptual Map,' unpublished paper, Carleton University 1991; Glen Toner, 'Whence and Whither: ENGOs, Business, and the Environment,' Carleton University, Centre for the Study of Business–Government–NGO Relations, Working Paper 1, 1991; Jeremy Wilson, 'Green Lobbies: Pressure Groups and Environmental Policy,' in Boardman, ed., *Canadian Environmental Policy*, 109–25; and Gregory Filyk and Ray Coté, 'Pressures from Inside: Advisory Groups and the Canadian Environmental Community,' in ibid, 60–82.

12 See Paul Pross, *Group Politics and Public Policy*, 2nd ed. (Toronto: Oxford University Press 1991); William S. Coleman and Grace Skogstad, eds., *Organized Interests and Public Policy* (Toronto: Copp-Clark 1990); and Michael Atkinson, ed., *Governing Canada: Institutions and Policy* (Toronto: Harcourt, Brace Jovanovich 1993).

13 See G. Bruce Doern, 'Canadian Policy Studies as Art, Craft and Science,' Paper presented to the Canadian Political Science Association Meetings, Carleton University, Ottawa, 6 June 1993.

14 The analysis of policy sectors and specific organizational resources draws from several elements of institutional political analysis. See Thomas Conway, 'The Policy Sector Approach to the Study of Public Policy: The Case of Federal Environmental Policy,' Unpublished paper, Carleton University, 1993; Thomas Conway, 'The Marginalization of the Department of the Environment: Environmental Policy 1971–1988,' Doctoral dissertation, Carleton University, 1992. See also Kenneth J. Benson, 'A Framework for Policy Analysis,' in David L. Roger and David A. Whetton, eds., *Interorganizational Coordination: Theory, Research and Implementation* (Des Moines: Iowa State University Press 1982), 137–76; G. Bruce Doern and Richard Phidd, *Canadian Public Policy: Ideas, Structure, Process*, 2d ed. (Toronto: Nelson Canada 1992); and Walter W. Powell and Paul J. DiMaggio, eds., *The New Institutionalism in Organizational Analysis* (Chicago: University of Chicago Press 1991).

15 See Donald Dewees, *The Regulation of Quality: Products, Services, Workplaces,*

and the Environment (Toronto: Butterworths 1983); Walter Block, ed., *Economics and the Environment: A Reconciliation* (Vancouver: Fraser Institute 1990); and David Pearce and R.K. Turner, *The Economics of Natural Resources and the Environment* (London: Harvester Wheatsheaf 1990).

16 See Peter Nemetz, 'Federal Environmental Regulation in Canada,' *Natural Resources Journal* 26 (1986), 551–608, and G. Bruce Doern, ed., *Getting It Green: Cases Studies in Canadian Environmental Regulation* (Toronto: C.D. Howe Institute 1990).

17 See Government of Canada, *Canada's Green Plan* (Ottawa: Minister of Supply and Services 1990).

18 Ibid.

19 A number of books and articles have analysed environmental values and ideas from different perspectives. See the following examples: Block, ed., *Economics and the Environment*; Thomas Conway, 'Challenges Facing Canadian Environmental Groups in the 1990s,' in Alain Gagnon and Brian Tanguay, eds., *Democracy with Justice: Essays in Honour of Khayyam Zev Paltiel* (Ottawa: Carleton University Press 1993), and 'The Concept of Sustainable Development,' in Donald Rowat, ed., *The Crisis of Environmental Degradation* (Ottawa: Carleton University Press 1989); Janet Foster, *Working for Wildlife: The Beginning of Preservation in Canada* (Toronto: University of Toronto Press 1978); Jacobs, *The Green Economy,* ch. 5; David Orfald and Robert Gibson, 'The Conserver Society Idea: A History with Questions,' *Alternatives* 12/3-4 (Spring/Summer 1985, 37–43; Paehlke, *Environmentalism and the Future of Progressive Politics* David Pepper, *The Roots of Modern Environmentalism* (London: Croom Helm 1984); and Michael Redclift, *Sustainable Development: Exploring the Contradictions* (New York: Methuen 1987).

20 See G. Bruce Doern, 'From Sectoral to Macro Green Governance: The Canadian Department of the Environment as an Aspiring Central Agency,' *Governance* 6/2 (April 1993), 172–93.

21 See George Hoberg, *Pluralism by Design: Environmental Policy and the American Regulatory State* (New York: Praeger 1992); Richard Harris and Sydney M. Milkis, *The Politics of Regulatory Change: A Tale of Two Agencies* (New York: Oxford University Press 1989); and David Vogel, *National Styles of Regulation* (Ithaca, NY: Cornell University Press 1986). For accounts of other OECD country environmental organizations, see Albert Weale, *The New Politics of Pollution* (Manchester: Manchester University Press 1992); Stephen Wilks, 'Administrative Culture and Policy Making in the Department of the Environment,' *Public Policy and Administration* 2/1 (Spring 1987), 25–41; and Jan van der Straaten, 'The Dutch National Environmental Policy Plan: To Choose or to Lose,' *Environmental Politics* 1/1 (Spring 1992), 45–71.

CHAPTER 1

1 See Michael Whittington, 'The Department of the Environment,' in G. Bruce Doern, ed., *Spending Tax Dollars* (Ottawa: School of Public Administration, Carleton University, 1980), 99.
2 Canada's environment department is by no means the only environmental agency to suffer from these classic dilemmas. On the U.K. situation, see Stephen Wilks, 'Administrative Culture and Policy Making in the Department of the Environment,' *Public Policy and Administration* 2/1 (Spring 1987), 25–41.
3 See *Speech from the Throne*, Prime Minister's Office, 8 October 1970, 1.
4 These quotations come from a series of briefing charts put together within the Prime Minister's Office (PMO). The charts set out, in graphic form, options and recommendations for the new department.
5 See Treasury Board, 'The Role and Functions of the Department of the Environment and Renewable Resources: A Conceptual Framework,' Planning Branch of the Treasury Board Secretariat, November 1970.
6 Ibid, 1.
7 Ibid.
8 Ibid, 2.
9 See Environment Canada, 'The Department of the Environment,' Lands Directorate, 1970.
10 Canadian Bar Association, '1971 CBA Resolutions' (Mimeo, 1971).
11 See Canadian Law and The Environment Workshop, *Report*, University of Manitoba 1971.
12 See Donella H. Meadows et al., *The Limits to Growth: A Report for the Club of Rome's Project on the Predicament of Mankind* (New York: Universe Books 1972).
13 See House of Commons, *Debates*, 9 October 1970.
14 These other components were found within what is now the Environmental Conservation and Protection side of the department. Inland Waters, for example, was more interested in considering the needs of all users, including those users causing pollution, and finding a way to balance those needs while ensuring the long-term viability of ecosystems. The EPS adopted the 'prevention at source' and 'best practical technology' positions originating with the strong Fisheries influence within that service. But the EPS had to deal with the reality of actually trying to implement pollution-control policies and programs. What they found, of course, was that negotiation with polluters, with all the limitations that implied, was often unavoidable.
15 See Environment Canada, 'Department of Environment: Preliminary Statement of Goals for 1980,' 14 January 1972.

16 Ibid, 1.
17 Ibid, 2.
18 See Environment Canada, 'Preliminary Report, Inhouse Operations: Department of the Environment,' 24 June 1974.
19 Memorandum from Assistant Deputy Minister Parks Canada to the Deputy Minister of Environment Canada, 28 March 1985.
20 Ibid, 2.
21 Ibid.
22 Ministers such as Jack Davis and Romeo LeBlanc had always neglected the Forestry Service. Their concern was for the Fisheries, and they determined that Forestry was a marginal federal government concern in relation to the provinces. However, in early 1979 the Trudeau government was becoming more interested in showing a higher federal government profile in several areas of national importance. The profile of Forestry was raised even farther in the early 1980s under Trudeau, and again after 1984 with the Mulroney Conservative government.
23 In many ways these conflicts were among the most virulent encountered within the department. Several interviews revealed that they hit at the core of the values held by most of the officials who could be attracted to an environment department. Pesticide use and forestry-exploitation rates were fundamental issues that had major structural consequences for the economy and ecological consequences for the natural environment.
24 See Environment Canada, 'Preliminary Report, Inhouse Operations,'
25 See G. Bruce Doern and V. Seymour Wilson, *Regulating Herbicides in the Canadian Forestry Industry: Issues and Problems* (Toronto: Canadian Council of Resource and Environmental Ministers 1986).
26 See Environment Canada, 'Senior Assistant Deputy Ministers Review of the Environmental Management Services Activities,' dated 1974.
27 See Environment Canada, 'Preliminary Statement of Goals.'
28 See Environment Canada, 'Review of the Environmental Management Service.'
29 See Environment Canada, 'Preliminary Statement of Goals.'
30 See ibid.
31 See Memorandum from Assistant Deputy Minister to Deputy Minister, 'Situation Report AES 1980/81,' 17 August 1981.
32 Ibid, 1–2.
33 See Atmospheric Environment Service, 'Strategic Plan: Challenges – 1990s and Beyond,' August 1987, 1.
34 Ibid, 11.

CHAPTER 2

1 This chapter relies on both interviews with several DOE ministers and federal and provincial officials as well as numerous internal memoranda and documents. For other valuable unpublished work on the DOE's first decade, see O.P. Dwividi and R. Brian Woodrow, 'Canada's First Environmental Decade 1971–1981,' unpublished paper, University of Guelph, December 1981.
2 See George Hoberg, Kathryn Harrison, and Karin Albert, 'The Politics of Canada's Green Plan,' Paper presented at the Canadian Political Science Association Meetings, Ottawa, 7 June 1993, 24–5.
3 Department of the Environment, 'The Montebello Goals,' 5 April 1971.
4 Ibid, 1.
5 Department of the Environment, 'A Ten Year Action Plan Environment Canada 1975–1985, Unpublished paper, Ottawa, February 1974.
6 Ibid, 5–6.
7 See G. Bruce Doern and Peter Aucoin, eds., *Public Policy in Canada: Structure and Organization* (Toronto: Macmillan of Canada 1979), ch. 1.
8 See Draft White Paper on the Environment, Draft No. 1, 12 October 1978.
9 See Department of the Environment, 'Major Priority Environmental Issues for the Eighties,' Ottawa, 7 December 1979.
10 See Environment Canada, '1984–1989 Strategic Plan,' Ottawa, October 1983.
11 Ibid, 10.
12 Ibid, 11.
13 See Study Team Report to the Task Force on Program Review, *Improved Program Review: Environment* (Ottawa: Minister of Supply and Services 1986).
14 Ibid, 11.
15 See Environment Canada, *Departmental Briefing Book*, Vol. 1 (Departmental Overview), Ottawa, March 1987.
16 Ibid, 1–13.
17 Ibid.

CHAPTER 3

1 See G. Bruce Doern and Richard W. Phidd, *Canadian Public Policy: Ideas, Structure, Process* (Toronto: Nelson Canada 1983), ch.10.
2 See François Bregha, *Bob Blair's Pipeline* (Toronto: Lorimer 1980).
3 See T. Fenge and Graham Smith, 'Reforming the Federal Environmental Assessment and Review Process,' *Canadian Public Policy* 12 (1986), 596–605.

4 See Stephen Clarkson, *Canada and The Reagan Challenge*, 2nd ed. (Toronto: McClelland and Stewart 1986).

5 See World Commission on Environment and Development, *Our Common Future* (Oxford: Oxford University Press 1987).

6 See Michael Whittington, 'The Department of the Environment,' in G. Bruce Doern, ed., *Spending Tax Dollars* (Ottawa: School of Public Administration, Carleton University, 1980), ch. 7.

7 See Memorandum from Deputy Minister to Minister, 'Financial Situation of the Department,' 28 June 1979.

8 Ibid, 2.

9 See Memorandum from Deputy Minister to Minister, 'Program Review,' 16 August 1979, 1.

10 For various ministerial defences of these cuts, see 'Notes for Speech by Suzanne Blais-Grenier, Minister of the Environment, Before the Standing Committee on Fisheries and Forestry,' Ottawa, 16 May 1985, and 'Notes for Statement by the Honourable Tom McMillan, Minister of the Environment Before the Standing Committee on Environment, and Forestry Regarding the Main Estimates for 1987–88,' Ottawa, undated.

11 See Allan Maslove, G. Bruce Doern, and Michael Prince, *Federal and Provincial Budgeting* (Toronto: University of Toronto Press 1986), ch. 6; and Douglas A. Smith, 'Defining the Agenda for Environmental Protection,' in Katherine A. Graham, ed., *How Ottawa Spends: 1990–91* (Ottawa: Carleton University Press 1990), 116–21.

12 The department was then called the Department of Indian and Northern Affairs (DINA), but throughout this section we use its present name, the Department of Indian Affairs and Northern Development (DIAND).

13 See Environment Canada, 'Briefing Notes for the Deputy Minister: Indian and Northern Affairs,' 4 July 1975.

14 Ibid.

15 Ibid.

16 See G. Bruce Doern and Glen Toner, *The Politics of Energy* (Toronto: Methuen 1985).

17 See G. Bruce Doern, ed., *The Environmental Imperative* (Toronto: C.D. Howe Institute 1990), ch. 1.

18 See Glen Toner, 'Stardust: The Tory Energy Program,' in Michael J. Prince, ed., *How Ottawa Spends: 1986–87* (Toronto: Methuen 1986), 119–48.

19 See Glen Toner, 'Whence and Whither: ENGOs, Business and the Environment,' Carleton University, Centre for the Study of Business–Government–NGO Relations, Working Paper No. 1, 1991.

20 See G. Bruce Doern, *Government Intervention in the Nuclear Industry* (Montreal: Institute for Research on Public Policy 1980).

21 See Donald Savoie, *Regional Economic Development: Canada's Search for Solutions* (Toronto: University of Toronto Press 1986); and G. Bruce Doern, 'The Department of Industry, Science and Technology: Is There Industrial Policy After Free Trade?,' in Katherine A. Graham, ed., *How Ottawa Spends: 1990–91* (Ottawa: Carleton University Press 1990), 49–72.

22 See K.E. DeSilva, *The Pulp and Paper Modernization Program: An Assessment* (Ottawa: Economic Council of Canada 1988).

23 See Letter from Deputy Minister of Regional Industrial Expansion to Deputy Minister of Environment, 19 August 1986, 2.

24 Ibid.

25 See Industry, Science and Technology Canada, *Environmental Industries Sector Initiative Workplan for 1989–90* (Ottawa, 1989). See also Paul M. Brown, 'Target or Participant? The Hatching of Environmental Industry Policy,' in Robert Boardman, ed., *Canadian Environmental Policy: Ecosystems, Politics and Process* (Toronto: Oxford University Press 1992).

26 See Office of the Science Advisor, *Maintenance and Utilization of Science in the Department of the Environment*, September 1983, 2.

27 See Kenneth Hare, 'Environmental Uncertainty: Science and the Greenhouse Effect,' in Doern, ed., *The Environmental Imperative*, 19–34.

28 Quoted in Office of the Science Advisor, *Maintenance and Utilization of Science*, 7.

29 Ibid, 2.

30 Ibid, 5.

31 See Department of the Environment, *Policy Respecting Science and Technology*, 20 October 1984.

32 For assessments of the evolving climate and politics of federal science policies in the 1970s and 1980s, see G. Bruce Doern, *Science and Politics in Canada* (Montreal and Kingston: McGill-Queen's University Press 1972); G. Bruce Doern, *The Peripheral Nature of Scientific and Technological Controversy in Federal Policy Formation* (Ottawa: Science Council of Canada 1981); and Doern, 'The Department of Industry, Science and Technology,'

33 See Canada, *Canada's Green Plan* (Ottawa: Minister of Supply and Services 1990).

34 See Glen Toner and Bruce Doern, 'Five Imperatives in the Formation of Green Plans: The Canadian Case,' *Environmental Politics* (forthcoming, 1994).

35 See George Hoberg, Kathryn Harrison, and Karin Albert, 'The Politics of Canada's Green Plan,' Paper presented at the Canadian Political Science Association Meetings, Ottawa, 7 June 1993, 17–18.

CHAPTER 4

1 See Grace Skogstad and Paul Kopas, 'Environmental Policy in a Federal System: Ottawa and the Provinces,' in Robert Boardman, ed., *Canadian Environmental Policy: Ecosystems, Politics and Process* (Toronto: Oxford University
Press 1992), 43–59.
2 See Norman C. Bonsor, 'Water Pollution and the Canadian Pulp and Paper
Industry,' in G. Bruce Doern, ed., *Getting It Green: Case Studies in Canadian
Environmental Regulation* (Toronto: C.D. Howe Institute 1990), 155–87.
3 On the pulp-and-paper sector, see William F. Sinclair, *Controlling Pollution
from Canadian Pulp and Paper Manufacturers: A Federal Perspective* (Ottawa:
Environment Canada 1988); and Bonsor, 'Water Pollution and the Canadian
Pulp and Paper Industry.'
4 See D.G. Kelley, 'Discussion Document,' Federal-Provincial Committee on
Air Quality, March 1990.
5 See G. Bruce Doern, *Regulatory and Jurisdictional Issues in the Regulation of Hazardous Substances* (Ottawa: Science Council of Canada 1977).
6 Kelly, 'Discussion Document,' 8.
7 See Allistaire R. Lucas, 'Harmonization of Federal and Provincial Environmental Policies: The Changing Legal and Policy Framework,' in J. Owen
Saunders, ed., *Managing Natural Resources in a Federal State* (Agincourt: Car-
swell 1986).
8 See Allaistaire R. Lucas, 'The New Environmental Law,' in Ronald L. Watts
and Douglas M. Brown, eds., *Canada: The State of the Federation 1989* (Kingston: Institute of Intergovernmental Relations, Queen's University, 1989),
167–92.
9 See Environmental Protection Service and Corporate Planning Group, 'Discussion Paper on the Future of the Federal-Provincial Accords for the Protec-
tions and Enhancement of Environmental Quality,' Ottawa, 11 January 1983.
10 Ibid, 2.
11 Quoted in L. Francis, 'Federal-Provincial Relations within Environment
Canada: An Overview,' Intergovernmental Affairs Directorate, March
1982, 9.
12 See G. Bruce Doern, *Green Diplomacy* (Toronto: C.D. Howe Institute 1993), 63.

CHAPTER 5

1 See Paul Pross, *Group Politics and Public Policy* (Toronto: Oxford University
Press 1986), and William D. Coleman and Grace Skogstad, eds., *Policy Communities and Public Policy in Canada* (Toronto: Copp Clark Pitman 1990).

2 See Mancur Olson, *The Logic of Collective Action* (Cambridge, MA: Harvard University Press 1971).

3 See Jeremy Wilson, 'Green Lobbies: Pressure Groups and Environmental Policy,' in Robert Boardman, ed., *Canadian Environmental Policy: Ecosystems, Politics and Process* (Toronto: Oxford University Press 1992), 109–25.

4 See Thomas Conway, 'Challenges Facing Canadian Environmental Groups in the 1990s,' in Alain Gagnon and Brian Tanguay, eds., *Democracy with Justice: Essays in Honour of Khayyam Zev Paltiel* (Ottawa: Carleton University Press 1992).

5 Ibid.

6 See G. Bruce Doern, Michael Prince, and Garth McNaughton, *Living with Contradictions: Health and Safety Regulation in Ontario* (Toronto: Royal Commission in Asbestos 1982).

7 See Glen Toner, 'The Canadian Environmental Movement: A Conceptual Map,' Unpublished paper, Carleton University 1991.

8 See Robert Paehlke, *Environmentalism and the Future of Progressive Politics* (New Haven, CT: Yale University Press 1989), ch. 2.

9 See Janet Foster, *Working for Wildlife: The Beginning of Preservation in Canada* (Toronto: University of Toronto Press 1978).

10 Science Council of Canada, *Canada as a Conserver Society* (Ottawa, 1977), and Donella H. Meadows et al. *The Limits to Growth: A Report for the Club of Rome's Project on the Predicament of Mankind* (New York: Universe Books 1972).

11 See Conway, 'Challenges,' and Toner, 'The Canadian Environmental Movement.'

12 See Glen Toner, 'Whence and Whither: ENGO, Business and the Environment,' Carleton University, Centre for the Study of Business–Government–NGO Relations, Working Paper No. 1, 1991, 3.

13 See Conway, 'Challenges,' and Toner, 'Whence and Whither.'

14 See Toner, 'The Canadian Environmental Movement,' 5.

15 See ibid, 4.

16 See Anna Bramwell, *Ecology in the 20th Century* (New Haven, CT: Yale University Press 1989).

17 Ibid, 6–7, and Bill Devall and Gerge Sessions, *Deep Ecology: Living As If Nature Mattered* (Salt Lake City: Peregrine Smith Books 1985).

18 Toner, 'The Canadian Environmental Movement,' 10.

19 See Department of the Environment, 'Funding Policies for Citizen's Groups,' May 1975, 3.

20 See G. Bruce Doern, ed., *The Environmental Imperative* (Toronto: C.D. Howe Institute 1990), 10.

21 See Donald Chant, 'A Decade of Environmental Concern: A Retrospective and Prospective,' *Alternatives*, 10/1 (1981), 3–6.
22 Department of the Environment, 'Funding Policies,' 4.
23 For further analysis, see Gregory Filyk and Ray Coté, 'Pressures from Inside: Advisory Groups and the Canadian Environmental Community,' in Boardman, ed., *Canadian Environmental Policy*, 60–82.
24 See Department of the Environment, 'Briefing Note on Canadian Coalition on Acid Rain,' July 1982.
25 See William D. Coleman, *Business and Politics: A Study of Collective Action* (Montreal and Kingston: McGill-Queen's University Press 1988).
26 See Benoît Laplante, 'Environmental Regulation: Performance and Design Standards,' in G. Bruce Doern, ed., *Getting It Green: Case Studies in Canadian Environmental Regulation* (Toronto: C.D. Howe Institute 1990), 59–88.
27 See Kernaghan Webb, 'Between Rocks and Hard Places: Bureaucrats, the Law and Pollution Control,' *Alternatives* 14 (May/June 1987), 4–14.
28 See, for example, Environment Canada, *Status Report on Compliance with Secondary Lead Smelter Regulations – 1984* (Ottawa: Environment Canada EPS, June 1985), and Environment Canada, *Status Report on Water Pollution Control in the Canadian Metal Mining Industry – 1986* (Ottawa: Environment Canada EPS, May 1988).
29 See Economic Council of Canada, *Responsible Regulation* (Ottawa: Minister of Supply and Services 1979).
30 See Toner, 'The Canadian Environmental Movement,' 9–10.
31 Ibid, 9.
32 See Department of the Environment, 'The Environment–Economy Thrust: A Corporate Planning Group Perspective,' August 1987.
33 See G. Bruce Doern, 'Regulations and Incentives: The NOx–VOCs Case,' in Doern, ed., *Getting It Green*, 89–110.
34 See Allan Howatson, *Business and the Environment: Economic Benefits from Environmental Improvements* (Ottawa: Conference Board of Canada 1991).
35 See Toner, 'The Canadian Environmental Movement,' *passim*.
36 See Michael Howlett, 'The Round Table Experience: Representation and Legitimacy in Canadian Environmental Policy Making,' *Queen's Quarterly* 97/4 (Winter 1990), 580–601.
37 See 'Industry, Lobbyists Join Forces,' *The Ottawa Citizen*, 28 September 1991, 5.
38 See Alan C. Cairns, 'Constitutional Change and the Three Equalities: Citizens, Provinces, Two Nations,' Paper presented to Symposium on Canada's Constitutional Options. Business Council on National Issues, Toronto, 16 January 1991.

39 See Michael Whittington, 'The Department of the Environment,' in G. Bruce Doern, ed., *Spending Tax Dollars* (Ottawa: School of Public Administration, Carleton University 1980), ch. 5.
40 See Department of Environment, 'Media Trends,' July 1979, 1.
41 See Canadian Trend Report, Summer 1983, (Montreal, 1983), 80.
42 See Department of Environment, Memorandum, 'Public Polling Views – Summer 1986,' Ottawa, 22 September 1986.
43 See Department of Environment, Undated Memorandum, 'Decima's Summer 1989 Survey,' Ottawa, 1989.
44 See George Hoberg, Kathryn Harrison and Karin Albert, 'The Politics of Canada's Green Plan,' Paper presented at the Canadian Political Science Association Meetings, Ottawa, 7 June 1993, 12–14, and G. Bruce Doern, *The Federal Green Plan: Assessing the Prequel* (Toronto: C.D. Howe Institute 1990).

CHAPTER 6

1 See Patrick Kyba, 'International Environmental Relations: Twenty Years of Canadian Involvement,' Unpublished paper, University of Guelph, 1990, and Robert Boardman, 'The Multilateral Dimension: Canada in the International System,' in Robert Boardman, ed., *Canadian Environmental Policy: Ecosystems, Politics and Process* (Toronto: Oxford University Press 1992), 224–45.
2 See George Hoberg, 'Sleeping with an Elephant: The American Influence on Canadian Environmental Regulation,' *Journal of Public Policy* 11 (1991), 107–31, and Don Munton, 'Dependence and Interdependence in Transboundary Environmental Relations,' *International Journal* 36 (1980–1), 139–84.
3 See L.K. Caldwell, *International Environmental Policy,* 2d ed. (Duram, NC: Duke University Press 1990), and Gareth Porter and Janet Welsh Brown, *Global Environmental Politics* (Boulder, CO: Westview Press 1991).
4 See Stephen H. Schneider, *Global Warming* (New York: Vintage Books 1990), 218–37, and G. Bruce Doern, 'Regulations and Incentives: The NOx–VOCs Case,' in G. Bruce Doern, ed., *Getting It Green: Case Studies in Canadian Environmental Regulation* (Toronto: C.D. Howe Institute 1990), 89–110.
5 See Michael Jacobs, *The Green Economy* (London: Pluto Press 1992), and Robert Goodland, H. Daly, S. El Serafy, and B. von Droste, *Environmentally Sustainable Economic Development: Building on Brundtland* (Paris: UNESCO 1991)
6 See Ray Robinson, 'Notes for an Address,' Paper presented at Seventh Symposium on Statistics and the Environment, National Academy of Sciences, Washington, DC, 4 October 1982, 3–6.
7 See Environment Canada, 'Situation Report 74/3,' Liaison and Coordination

Directorate, 22 May 1974, and 'Situation Report 77/6,' Liaison and Coordination Directorate, 1 June 1977.

8 See Robinson, 'Notes for an Address,' 22–30.

9 See George Hoberg, 'Comparing Canadian Performance in Environmental Policy,' in Boardman, ed., *Canadian Environmental Policy,* 246–62.

10 See Richard A. Harris and Sidney M. Milkis, *The Politics of Regulatory Change* (New York: Oxford University Press 1989), ch. 6, and George Hoberg, *Pluralism by Design: Environmental Policy and the American Regulatory State* (New York: Praeger 1992).

11 See Environment Canada, *Inventory of International Activities* (Ottawa: Supply and Services Canada 1986).

12 See Caldwell, *International Environmental Policy,* ch. 3 and 4.

13 See International Affairs Branch, Environment Canada, 'International Environmental Issues: A Status Report,' January 1989.

14 Ibid, 1.

15 Ibid.

16 See Doern, 'Regulations and Incentives,' 94–6.

17 See Doern, 'Regulations and Incentives.'

18 See G. Bruce Doern, *Green Diplomacy* (Toronto: C.D. Howe Institute 1993), ch. 2. See also G. Victor Buxton, 'UNCED and Lessons from the Montreal Protocol,' *Law and Contemporary Problems* forthcoming issue (1994).

19 See Elizabeth P. Barat Brown, 'Building a Monitoring and Compliance Regime under the Montreal Protocol,' *Yale Journal of International Law* 16/2 (1990), 520–70, and R.E. Benedick, *Ozone Diplomacy: New Directions in Safeguarding the Planet* (Cambridge, MA: Harvard University Press 1991).

20 See Doern, *Green Diplomacy,* ch. 4, and Andrew Hurrell and Benedict Kingsbury, eds., *The International Politics of the Environment* (Oxford: Clarendon Press 1992).

21 See Fen Osler Hampson, 'Climate Change and Global Warming: The Elusive Search for an International Convention,' Occasional Paper Series, Maxwell School of Public Affairs, Syracuse University, 6 December 1990; and William A. Nitze, *The Greenhouse Effect: Formulating a Convention* (London: Royal Institute for International Affairs 1990).

22 See Jessica Tuchman Mathews, 'Redefining Security,' *Foreign Affairs* 68/2 (Spring 1989), 162–77.

23 Fen Osler Hampson, 'New Wine in an Old Bottle? The International Politics of the Environment,' Paper presented to the Conference on the New World Situation and the Future of Peace, Hobart and William Smith Colleges, Geneva, New York, 27 April 1991, 1.

24 See Robert E. Goodin, 'The High Ground Is Green,' *Environmental Politics* 1/1

(Spring 1992), 1–8, and Andrew Simms, 'If Not Then, When? Non-Governmental Organizations and the Earth Summit Process,' *Environmental Politics* 2/1 (Spring 1993), 94–101.

25 See Caldwell, *International Environmental Policy,* 3.

26 Ibid.

27 See Jim MacNeill and Robert Munro, 'Environment 1972–92: From the Margin to the Mainstream,' *Ecodecision* 1 (1991), 19–23.

28 World Commission on Environment and Development, *Our Common Future* (Oxford: Oxford University Press 1987), 43.

29 See Doern, *Green Diplomacy,* Ch. 3; Andrew Hurrell, "The 1992 Earth Summit: Funding Mechanisms and Environmental Institutions,' *Environmental Politics* 1/4 (Winter 1992), 273–9; and Jean-Guy Vaillancourt, 'UNCED and 92 Global Forum: The Rio Summits,' *Delta* 3/2, (Fall 1992), 1–5.

30 See G. Bruce Doern, 'International Institutional Issues In the Convergence of Investment, Trade and Environmental Policies,' Paper prepared for Investment Canada, Ottawa, July 1993.

31 See *United States – Restrictions in Imports of Tuna,* Report of the GATT Panel, 3 September 1991.

32 For a range of views, see Kym Anderson and Richard Blackhurst, eds., *The Greening of World Trade Issues* (London: Harvester Wheatsheaf 1992); John Whalley, 'The Interface between Environmental and Trade Policies,' *The Economic Journal* 101 (March 1991), 180–9; and Business Council on National Issues, 'International Trade: The Environmental Dimension,' *Research Bulletin,* April 1993.

33 See Baratt Brown, 'Building a Monitoring and Compliance Regime.'

34 See Government of Canada, *North American Free Trade: The NAFTA Manual* (Ottawa, August 1992). This is not the actual NAFTA agreement but rather a summary of key points.

35 On these trade remedy issues, see Thomas S. Boddez and Michael J. Trebilock, *Unfinished Business: Reforming Trade Laws in North America* (Toronto: C. D. Howe Institute 1993).

36 Ibid, 2.

37 Ibid, 1.

CHAPTER 7

1 See Chris Park, *Acid Rain: Rhetoric and Reality* (New York: Methuen/Routledge 1989), and D. Drablos and A. Tollan, eds., *Ecological Impact of Acid Precipitation,* Proceedings of conference held in Sandefjord, Norway, 11–14 March 1980 (Oslo, October 1980).

2 See G. Bruce Doern, *The Politics of Risk: The Identification of Hazardous Substances in Canada* (Toronto: Royal Commission on Asbestos 1982).

3 See Organization for Economic Cooperation and Development (OECD), *The OECD Programmme on Long Range Transport of Air Pollutants* (Paris, 1977).

4 See Science Council of Canada, *Policies and Poisons* (Ottawa, 1977).

5 See DOE Paper, 'Acid Rain Briefing Package for Mr Roberts' Visit to Washington,' 18 April 1980.

6 See Robert B. Gibson, 'Out of Control and Beyond Understanding: Acid Rain as a Political Dilemma,' in Robert Paehlke and Douglas Torgerson, eds., *Managing Leviathan: Environmental Politics and the Administrative State* (Peterborough: Broadview Press 1990), 243–82.

7 On U.S. dynamics and issues, see James L. Regens and Robert W. Rycroft, *The Acid Rain Controversy* (Pittsburgh: University of Pittsburg Press 1988).

8 Environental attitudes in the Reagan Administration were tied to larger views about limiting the role of government and reducing regulatory intervention. See Richard A. Harris and Sidney M. Milkis, *The Politics of Regulatory Change: A Tale of Two Agencies* (New York: Oxford University Press 1989).

9 See Raymond M. Robinson, 'Notes for an Address,' Seventh Symposium on Statistics and the Environment, National Academy of Sciences, 4–5 October 1982, Washinton, DC, 23.

10 See 'Proceedings, International Conference of Ministers on Acid Rain,' Ottawa, March 1984.

11 On the European acid-rain story see S. Boehmer-Christiansen, *Acid Politics* (London: Macmillan 1990).

12 See David M. Newbery, 'Acid Rain,' *Economic Policy* 5/2 (October 1990), 297–346.

13 See Stephen Clarkson, *Canada and the Reagan Challenge*, 2d ed. (Toronto: Lorimer 1985).

14 See Donald Dewees, 'The Regulation of Sulphur Dioxide in Ontario,' in G. Bruce Doern, ed., *Getting It Green: Case Studies in Canadian Environmental Regulation* (Toronto: C.D. Howe Institute 1990), 129–54.

15 See Letter from Keith Norton to John Roberts, 30 October 1981, and DOE Memorandum from Deputy Minister to Minister, Undated but referring to the 30 October letter.

16 See G. Bruce Doern, Michael Prince, and Garth McNaughton, *Living with Contradictions: Health and Safety Regulation in Ontario* (Toronto: Royal Commission on Asbestos 1982).

17 See Canada, *Canada's Nonferrous Metals Industry: Nickel and Copper* (Ottawa: Minister of Supply and Services 1984), iii.

18 For some of the differences in each of the companies involved see, ibid, ch. 4, 5, and 6.
19 See DOE Memorandum from A.N. Manson to 'Distribution,' 2 October 1984.
20 See Drew Lewis and William Davis, *Joint Report of the Special Envoys on Acid Rain* (Ottawa: Minister of Supply and Services, January 1986).
21 See Richard E. Cohen, *Washington at Work: Back Rooms and Clean Air* (New York: Macmillan 1992).
22 In their account of the dynamics leading to a new Clean Air Act in the United States, Harris and Milkis *The Politics of Regulatory Change*, 270 point to Canadian pressure as a significant factor.
23 For example, in May 1984, a poll showed that 78 per cent of Canadians were either extremely or somewhat concerned about acid rain and that 54 per cent would willingly pay for an acid-rain clean-up. See *The Citizen* (Ottawa), 10 May 1984, 10.
24 See House of Commons, *Still Waters*. Report of the Sub-Committee on Acid Rain of the Standing Committee on Fisheries and Forestry (Ottawa: Minister of Supply and Services 1981).
25 See House of Commons, *Time Lost: A Demand For Action on Acid Rain*. Report of the Sub-Committee on Acid Rain of the Standing Committee on Fisheries and Forestry (Ottawa: Minister of Supply and Services 1984).

CHAPTER 8

1 See W.F. Lothian, *A History of Canada's National Parks*, vol. I (Ottawa: Parks Canada 1976).
2 See John Richards and Larry Pratt, *Prairie Capitalism* (Toronto: McClelland and Stewart 1979).
3 See Department of Environment, *Major Departmental Priorities for the Eighties* (Ottawa, 7 December 1979), 6.
4 See Task Force on Program Review, *Environment* (Ottawa: Supply and Services Canada 1985), 54.
5 Quoted in Ibid, 35.
6 See Department of the Environment, *Environment Canada 1984–1989 Strategic Plan*, 20 October 1983, 17.
7 See Task Force on Program Review, *Environment*, 54.
8 See CPER Management Consulting Inc., *Evaluation Framework of the National Historic Parks and Sites Evaluation of Parks Canada*, Prepared for Environment Canada, Sepember 1985.
9 Task Force on Program Review, *Environment*, 55.
10 Ibid, 56.

11 See G. Bruce Doern, *The Politics of Slow Progress: Federal Aboriginal Policy Processes* (Ottawa: Royal Commission on Aboriginal Peoples 1993).

12 Task Force on Program Review, *Environment*, 60.

13 See Bruce Downie and Bob Peart, eds., *Parks and Tourism: Progress or Prostitution* (Victoria: National and Provincial Parks Association of Canada, British Columbia Chapter, 1982).

14 See Elizabeth May, *Paradise Won: The Struggle for South Moresby* (Toronto: McClelland and Stewart 1990). This excellent account of the South Moresby story has been an invaluable source for this chapter. We have tried to put South Moresby more into the context of Parks Canada's own evolution, something which the May book does not do. Our additional perspective is largely based on interviews but also on some of the internal correspondence and documents at the DOE.

15 See May, *Paradise Won*, ch. 1–5.

16 See ibid, ch. 12.

17 See ibid, ch. 28.

18 See J.G. Nelson, *An External Perspective on Parks Canada Strategies, 1986–2001*, (Waterloo: University of Waterloo, January 1984).

19 See Minister of Environment's Task Force on Park Establishment, *Our Parks – Vision For the 21st Century*, University of Waterloo, June 1987. See also its briefer *Parks 2000: Vision for the 21st Century*, Univerity of Waterloo, June 1987.

20 Ibid, Nelson, *An External Perspective*, 4.

21 Ibid, 16.

22 See Environment Canada Parks Service, *State of the Parks 1990 Report* (Ottawa: Minister of Supply and Services 1991).

23 See Environment Canada Parks Service, *Environment Canada: The Canadian Parks Service Strategic Plan* (Ottawa: Minister of the Environment 1991), 8.

CHAPTER 9

1 See Robert V. Bartlett, 'Ecological Reason in Administration: Environmental Impact Assessment and Administrative Theory,' in Robert Paehlke and Douglas Torgerson, eds., *Managing Leviathan: Environmental Politics and the Administrative State* (Peterborough: Broadview Press 1990), pp. 81–96; T. Fenge and Graham Smith, 'Reforming the Federal Environmental Assessment and Review Process,' *Canadian Public Administration* 12 (1986), 596–605; and G. Bruce Doern, *Canadian Environmental Policy: Why Process Is Almost Everything* (Toronto: C.D. Howe Institute Commentary 19, 1990).

2 There are several good FEARO Occasional Papers that outline the EARP process. See David Marshall, 'Environmental Management and Impact Assess-

ment,' March 1985; Raymond Robinson, 'The Federal Role in Environmental Assessment,' April 1985; and W.J. Couch et al. 'Environmental Impact Assessment in Canada,' 1981.

3 Many studies have discussed the outward expansion of development into the Canadian hinterland. See Harold Innis, *Problems of Staple Production in Canada* (Toronto: Ryerson Press 1933), A.R.M. Lower, *The North American Assault on the Canadian Forests* (Toronto: Ryerson Press 1938); and Donald MacKay, *Heritage Lost: The Crisis in Canada's Forests* (Toronto: Macmillan 1985).

4 Quoted in Ken Clark, *Environment Canada Chronologies: 1970–1990* (Ottawa: Environment Canada, April 1991), Section H.

5 Ibid.

6 See George Hoberg, *Pluralism by Design: Environmental Policy and the American Regulatory State* (New York: Praeger 1992), and George Hoberg, 'Environmental Policy: Alternative Styles,' in Michael Atkinson, ed., *Governing Canada: Institutions and Policy* (Toronto: Harcourt Brace Jovanovich 1993), 307–42.

7 See F. Anderson, *NEPA and the Courts* (Baltimore, MD: Johns Hopkins University Press 1973); L.K. Caldwell, *Science and the National Environmental Policy Act: Redirecting Policy Through Procedural Reform* (Birmingham: University of Alabama Press 1982); and S. Taylor, *Making Bureaucracies Think: The Environmental Impact Statement Strategy of Administrative Reform* (Stanford, CA: Stanford University Press 1984).

8 Memo dated 4 September 1973.

9 Ibid, 1.

10 Ibid, 2.

11 Memo from Reed Logie to Robert Shaw, 25 April 1974, 1.

12 See Environment Canada, 'Environmental Evaluation Process,' 19 December 1975. It was sent by Gauthier in response to Seaborn's 2 October 1975 internal DOE initiative to institute new procedures for the application of the EARP.

13 See Environment Canada, 'Review of the Environmental Assessment and Review Process,' 7 April 1975.

14 Ibid, 1.

15 See Environment Canada, 'Amendments to the Environmental Assessment and Review Process: Submission to the Interdepartmental Committee on the Environment,' 14 January 1976.

16 See Environment Canada, 'Proposed Cabinet Memorandum Recommending Amendments to the Environmental Assessment and Review Process,' 12 November 1976. The document also included 'Briefing Notes for Minister on Memorandum to Cabinet.'

17 Ibid, 2.

18 Ibid.

19 See Environment Canada, 'Departmental Environmental Assessment Policy Review,' May 1991.

20 Ibid, 3.

21 Quoted in Clark, *Environment Canada Chronologies: 1970–1990*, Section H.

22 This confidential letter, sent by the prime minister, responded to Marchand's proposals regarding the separation of Fisheries from Environment and provided instruction on a range of issues regarding the second 'new' DOE.

23 Ibid, 2.

24 Ibid, 4.

25 See Environment Canada, 'EARP Review – Status Report,' 30 June 1983. An attached discussion document is titled 'The Federal Environmental Assessment and Review Process.'

26 Environment Canada, 'EARP Review,' 2.

27 Ibid, 5.

28 Ibid, 6.

29 See Canada, *Environmental Assessment and Review Process Guidelines Order* (1984), SOR/84-467.

30 See Environment Canada, 'Departmental Environmental Assessment Policy Review,' 2 May 1991.

31 See the Royal Commission on the Economic Union and Development Prospects for Canada (1985), *Report*, vol. 2 (Ottawa: Minister of Supply and Services 1985).

32 See Canadian Environmental Advisory Council, *Preparing For the 1990s: Environmental Assessment* (Ottawa: Minister of Supply and Services 1988).

33 See T.R. Schrecker, 'Of Invisible Beasts and the Public Interest: Environmental Cases and the Judicial System,' in Robert Boardman, ed., *Canadian Environmental Policy: Ecosystems, Politics and Process* (Toronto: Oxford University Press 1992).

34 See T.F. Schrecker, 'The Canadian Environmental Assessment Act: Tremulous Step Forward, or Retreat into Smoke and Mirrors,' *Canadian Environmental Law Reports* 5 (March 1991), 192.

35 Ibid.

36 See Hoberg, 'Environmental Policy,' 326.

37 See Environment Canada, 'Departmental Environmental Assessment Policy Review,' 2 May 1991.

38 See Doern, *Canadian Environmental Policy: Why Process is Almost Everything*, and G. Bruce Doern, 'From Sectoral to Macro Green Governance: The Canadian Department of the Environment as an Aspiring Central Agency,' *Governance* 6/2 (April 1993), 172–93.

CHAPTER 10

1 See also Thomas Conway, 'Taking Stock of the Traditional Regulatory Approach,' in G. Bruce Doern, ed., *Getting It Green: Case Studies in Canadian Environmental Regulations* (Toronto: C.D. Howe Institute 1990), 25–58.

2 See Paul R. Portney, ed., *Public Policies for Environmental Protection* (Washington: Resources for the Future 1990), and David Vogel, *National Styles of Regulation: Environmental Policy in Great Britain and the United States* (Ithaca, NY: Cornell University Press 1986).

3 See Benoît Laplante, 'Environmental Regulation: Performance and Design Standards,' in Doern, ed., *Getting It Green*, 59–88.

4 See Kernaghan Webb, *Pollution Control in Canada: The Regulatory Approach in the 1980s* (Ottawa: Law Reform Commission of Canada 1986); T.F. Schrecker, *The Political Economy of Environmental Hazards* (Ottawa: Law Reform Commission of Canada 1984); T.F. Schecker, 'Resisting Regulation: Environmental Policy and Corporate Power,' *Alternatives* 14/1 (1987), 9–21; Thomas Ilgen, 'Between Europe and America, Ottawa and the Provinces: Regulating Toxic Substances in Canada,' *Canadian Public Policy,* 11/3 (1985), 578–90, and G. Bruce Doern, *Regulatory and Jurisdictional Issues in the Regulation of Hazardous Substances in Canada* (Ottawa: Science Council of Canada 1977).

5 See Law Reform Commission, *Policing Pollution* (Ottawa, 1982), and John Swaigen and Gail Bunt, *Sentencing in Environmental Cases* (Ottawa: Law Reform Commission of Canada 1985).

6 This chronology of events was drawn from internal DOE discussion documents and briefing papers. See also Ken Clark, *Environment Canada Chronologies: 1970–1990* (Ottawa: DOE, April 1991).

7 This quotation appeared often in DOE internal documents. It became a key rallying cry regarding use of the Fisheries Act in environmental regulation.

8 This quotation was drawn from a confidential internal DOE discussion paper sent by Raymond Robinson, acting assistant deputy minister of the Environmental Protection Service, to Blair Seaborn, deputy minister. The document addresses the DOE's status as regulator upon the split of the Fisheries Service from the department. See 'Submission to the Senior Management Committee Environment Canada: Environmental Protection Strategy With Specific Focus On The Regulatory Aspects,' 30 March 1979.

9 Department of Fisheries and Forestry, 'Notes for a Speech in the House of Commons on the Speech from the Throne, Thursday, October 15, 1970.'

10 Ibid.

11 See Environment Canada, 'Second Draft Statement on Legislation,' 22 June 1972. This discussion document was sent to all members of the 'Cross Mis-

sion Task Force on Environmental Contaminants' from the 'Legal Sub-Committee,' which had been established to consider options for new legislation.

12 See Memo dated 30 March 1979 from Raymond Robinson to Deputy Minister Blair Seaborn.

13 See Environment Canada, 'Working Paper on Proposed *Environmental Contaminants Act* for Inter-departmental Discussion and Discussion with Provincial Authorities,' 18 April 1973.

14 Ibid, 3.

15 See Donald A. Chant and Ross H. Hall, 'Ecotoxicity in Canada,' discussion paper, Canadian Environmental Advisory Council, Ottawa, 1979, 5.

16 Ibid, 10.

17 It is very difficult to get reliable data on formal legal undertakings and prosecutions. This is an estimated provided by several DOE regulatory officials.

18 See Clark, *Environment Canada Chronologies, 1970–1990*, for a chronology of these developments.

19 See the Canadian Trend Report, *Environment*, for the last half of 1982, 93.

20 See the Canadian Trend Report, *Environment*, for the first half of 1983, 80.

21 Canada, Inquiry on Federal Water Policy, *Currents of Change* (Ottawa: Minister of Supply and Services 1985).

22 Quoted in Environment Canada, 'Ministerial Briefing Paper,' 5 February 1986.

23 Ibid.

24 See 'Notes for Remarks by The Honourable Tom MacMillan, P.C., M.P., Federal Minister of the Environment, to the Legislative Committee on Bill C-74, The Canadian Environmental Protection Act,' Ottawa, 25 November 1987.

25 See Environment Canada, *Canadian Environmental Protection Act*. For Years Ending March 1990, 1991, and 1992 (Ottawa: Minister of Supply and Services).

CHAPTER 11

1 See Herman Bakvis and Neil Nevitte, 'The Greening of the Canadian Electorate: Environmentalism, Ideology and Partisanship,' in Robert Boardman, ed., *Canadian Environmental Policy: Ecosystems, Politics and Process* (Toronto: Oxford University Press 1992), 144–63.

2 See Al Davidson and Tony Hodge, 'The Fate of the Great Lakes,' *Policy Options* 10/8 (October 1989), 19–26.

3 See Glen Toner, 'First Generation Green Plans: The Canadian Experience,' *Eco-Decision* 1 (1991); G. Bruce Doern, *Shades of Green: Gauging Canada's Green Plan* (Toronto: C.D. Howe Institute Commentary, 1991)' and George Hoberg,

Kathryn Harrison, and Karin Albert, 'The Politics of Canada's Green Plan,' Paper presented at the Canadian Political Science Association Meetings, Ottawa, 7 June 1993.

4 See G. Bruce Doern, *Canadian Environmental Policy: Why Process Is Almost Everything* (Toronto: C.D. Howe Institute Commentary, 1990).

5 See G. Bruce Doern and Richard Phidd, *Canadian Public Policy: Ideas, Structure Process*, 2d ed., (Toronto: Nelson Canada 1992), ch. 3, and Peter Hall, *The Political Power of Ideas: Keynesianism Across Ideas* (Princeton, NJ: Princeton University Press 1989).

6 See G. Bruce Doern, 'From Sectoral to Macro Green Governance: The Canadian Department of the Environment as an Aspiring Central Agency,' *Governance* 6/2 (April 1993), 172–93.

7 Michael Jacobs, *The Green Economy* (London: Pluto Press 1991), 59.

8 For further detailed analysis, see Doern, 'From Sectoral to Macro Green Governance.'

9 See G. Bruce Doern, ed., *Getting It Green: Case Studies in Canadian Environmental Regulation* (Toronto: C.D. Howe Institute 1990).

10 See Frances Cairncross, *Costing the Earth* (London: The Economist Books 1991).

11 See Richard Simeon, 'Globalization and the Canadian Nation State,' in G. Bruce Doern and Bryne Purchase, eds., *Canada at Risk? Canadian Public Policy in the 1990s* (Toronto: C.D. Howe Research Institute 1991), 46–58.

12 See Jacobs, *The Green Economy.*

13 See Doern and Phidd, *Canadian Public Policy,* ch. 3–7.

14 See Jacobs, *The Green Economy.*

15 See G. Bruce Doern, *Science and Politics in Canada* (Montreal and Kingston: McGill-Queens University Press 1972), and G. Bruce Doern, *The Peripheral Nature of Scientific and Technological Controversy in Federal Policy Formation* (Ottawa: Science Council of Canada 1981).

Selected Bibliography

BOOKS AND ARTICLES

Anderson, F. *NEPA in the Courts.* Baltimore, MD: Johns Hopkins University Press 1973.

Anderson, Kym, and Richard Blackhurst, eds. *The Greening of World Trade Issues.* London: Harvester Wheatsheaf 1992.

Atkinson, Michael, ed. *Governing Canada: Institutions and Policy.* Toronto: Harcourt Brace Jovanovich 1993.

Bakvis, Herman, and Neil Nevitte. 'The Greening of the Canadian Electorate: Environmentalism, Ideology and Partisanship.' In Robert Boardman, ed. *Canadian Environmental Policy.* Toronto: Oxford University Press 1992, 144–63.

Barat Brown, Elizabeth P. 'Building a Monitoring and Compliance Regime Under the Montreal Protocol.' *Yale Journal of International Law* 16/2 (1990), 520–70.

Benson, Kenneth J. 'A Framework For Policy Analysis.' In David L. Roger and David A. Whetton, eds. *Interorganizational Coordination: Theory, Research and Implementation.* Des Moines: Iowa State University Press 1982, 137–76.

Birnie, Patricia W., and Alan E. Boyle. *International Law and the Environment.* Oxford: Clarendon Press 1992.

Block, Walter, ed. *Economics and the Environment: A Reconciliation.* Vancouver: Fraser Institute 1990.

Boardman, Robert. *Canadian Environmental Policy: Ecosystems, Politics and Process.* Toronto: Oxford University Press 1992.

Bonsor, Norman C. 'Water Pollution and the Canadian Pulp and Paper Industry.' In G. Bruce Doern, ed. *Getting It Green: Case Studies in Canadian Environmental Regulation.* Toronto: C.D. Howe Institute 1990, 155–87.

Bonsor, Norman C., N. McCubbin, and J.B. Sprague. *Kraft Mill Effluent in Ontario*. Toronto: Ministry of the Environment 1988.

Bramwell, Anna. *Ecology in the Twentieth Century*. New Haven, CT: Yale University Press 1989.

Bregha, François. *Bob Blair's Pipeline*. Toronto: Lorimer 1980.

Bregha, Francois, J. Benedickson, D. Gamble, T. Shillington, and E. Weick. *The Integration of Environmental Considerations in Government Policy*. Ottawa: Rawson Academy of Aquatic Science 1990.

Brown, Paul M. 'Organizational Design as a Policy Instrument: Environment Canada in the Canadian Bureaucracy.' In Robert Boardman, ed. *Canadian Environmental Policy: Ecosystems, Politics and Process*. Toronto: Oxford University Press 1992, 24–42.

– 'Target or Participant? The Hatching of Environmental Industry Policy.' In Robert Boardman, ed. *Canadian Environmental Policy: Ecosystems, Politics and Process*. Toronto: Oxford University Press 1992, 164–78.

Cairncross, Frances. *Costing the Earth*. London: Economist Books 1991.

Caldwell, L.K. *International Environmental Policy*, 2d ed. Durham, NC: Duke University Press 1990.

– *Science and the National Environmental Policy Act: Redirecting Policy through Procedural Reform*. Birmingham: University of Alabama Press 1982.

Canada. *Canada's Green Plan*. Ottawa: Minister of Supply and Services 1990.

– *Canada's Green Plan: The First Year*. Ottawa: Minister of Supply and Services 1991.

– *Canada's Green Plan: The Second Year*. Ottawa: Minister of Supply and Services 1993.

Canadian Environmental Advisory Council. *Canada and Sustainable Development*. Ottawa: Minister of Supply and Services 1987.

– *Enforcement Practices of Environment Canada*. Ottawa: Minister of Supply and Services 1987.

– *Evaluating Environmental Impact Assessment*. Ottawa: Minister of Supply and Services 1988.

– *Examining Environment–Economy Linkages*. Ottawa: Minister of Supply and Services 1986.

– *Free Trade and the Environment*. Ottawa: Minister of Supply and Services 1986.

– *Preparing for the 1990s: Environmental Assessment: An Integral Part of Decision Making*. Ottawa: Minister of Supply and Services 1988.

– *Selling Canada's Environment Short: The Environmental Case against the Free Trade Deal*. Toronto, 1986.

Canadian Environmental Law Research Foundation. *Pollution and the Law – New Players – New Rules – New Ball Game*. Toronto, 1987.

Canadian Institute of Resources Law. *Environmental Law in the 1980s: A New Beginning.* Calgary, 1981.

Chant, Donald. 'A Decade of Environmental Concern: A Retrospective and Prospective.' *Alternatives* 10 (Spring-Summer 1981), 3–7.

Clarkson, Stephen. *Canada and the Reagan Challenge,* 2d ed. Toronto: Lorimer 1985.

Cohen, Richard E. *Washington at Work: Back Rooms and Clean Air.* New York: Macmillan 1992.

Coleman, William D. *Business and Politics: A Study of Collective Action.* Montreal and Kingston: McGill-Queen's University Press 1988.

Coleman, William D., and Grace Skogstad, eds. *Policy Communities and Public Policy in Canada.* Toronto: Copp Clark Pitman 1990.

Conway, Thomas. 'Challenges Facing Canadian Environmental Groups in the 1990s.' In Alain Gagnon and Brian Tanguay, eds. *Democracy with Justice: Essays in Honour of Khayyam Zev Paltiel.* Ottawa: Carleton University Press 1993, 159–73.

– 'The Marginalization of the Department of the Environment: Environment Policy 1971–1988.' Doctoral dissertation, Carleton University, Ottawa 1992.

– 'Taking Stock of the Traditional Regulatory Approach.' In G. Bruce Doern, ed. *Getting It Green: Case Studies in Canadian Environmental Regulation.* Toronto: C.D. Howe Institute 1990, 25–58.

Couch, William, ed. *Environmental Assessment in Canada: 1988 Summary of Current Practice.* Ottawa: Canadian Council of Resource and Environment Ministers 1988.

Davidson, Al, and Tony Hodge. 'The Fate of the Great Lakes.' *Policy Options* 10/8 (October 1989), 19–26.

DeSilva, K.E. *The Pulp and Paper Modernization Program: An Assessment.* Ottawa: Economic Council of Canada 1988.

Dewees, Donald. 'Instrument Choice in Environmental Policy.' *Economic Enquiry* 21/1 (January 1983), 53–71.

– *The Regulation of Quality: Products, Services, Workplaces, and the Environment.* Toronto: Butterworths 1983.

– 'The Regulation of Sulphur Dioxide in Ontario.' In G. Bruce Doern, ed. *Getting It Green: Case Studies in Canadian Environmental Regulation.* Toronto: C.D. Howe Institute 1990, 129–54.

Doern, G. Bruce. 'The Department of Industry, Science and Technology: Is There Industrial Policy After Free Trade?' In Katherine A. Graham, ed. *How Ottawa Spends: 1990–91.* Ottawa: Carleton University Press 1990, 49–92.

– 'From Sectoral to Macro Green Governance: The Canadian Department of the Environment as an Aspiring Central Agency.' *Governance* 6/2 (April 1993, 172–93.

– *Government Intervention in the Nuclear Industry.* Montreal: Institute For Research on Public Policy 1980.
– *Green Diplomacy.* Toronto: C.D. Howe Institute 1993.
– *The Peripheral Nature of Scientific and Technological Controversy in Federal Policy Formation.* Ottawa: Science Council of Canada 1981.
– *The Politics of Risk: The Identification of Hazardous Substances in Canada.* Toronto: Royal Commission on Asbestos 1982.
– 'Regulations and Incentives: The NOx–VOCs Case.' In G. Bruce Doern, ed. *Getting It Green: Case Studies in Canadian Environmental Regulation.* Toronto: C.D. Howe Institute 1990, 89–110.
– *Regulatory and Jurisdictional Issues in the Regulation of Hazardous Substances in Canada.* Ottawa: Science Council of Canada 1977.
– *Science and Politics in Canada.* Montreal and Kingston: McGill-Queen's University Press 1972.
Doern, G. Bruce, ed. *The Environmental Imperative.* Toronto: C.D. Howe Institute 1990.
– *Getting It Green: Case Studies in Canadian Environmental Regulation.* Toronto: C.D. Howe Institute 1990.
Doern, G. Bruce, and Peter Aucoin, eds. *Public Policy in Canada.* Toronto: Macmillan of Canada 1979.
Doern, G. Bruce, and Richard W. Phidd. *Canadian Public Policy: Ideas, Structure, Process,* 2d ed. Toronto: Nelson Canada 1992.
Doern, G. Bruce, Michael Prince, and Garth McNaughton. *Living with Contradictions: Health and Safety Regulation in Ontario.* Toronto: Royal Commission on Asbestos 1982.
Doern, G. Bruce, and Glen Toner. *The Politics of Energy.* Toronto: Methuen 1985.
Doern, G. Bruce, and V. Seymour Wilson. *Regulating Herbicides in the Canadian Forestry Industry: Issues and Problems.* Toronto: Canadian Council of Resource and Environmental Ministers 1986.
Downie, Bruce, and Bob Peart, eds. *Parks and Tourism: Progress or Prostitution?* Victoria: National and Provincial Parks Association of Canada, British Columbia Chapter, 1982.
Dwividi, O.P. 'The Canadian Government's Response to Environmental Concern.' In O.P. Dwividi, *Protecting the Environment: Issues and Choices – Canadian Perspectives.* Toronto: Copp Clark 1974.
– *Resources and the Environment: Policy Perspectives for Canada.* Toronto: McClelland and Stewart 1980.
Economic Council of Canada. *Responsible Regulation.* Ottawa: Minister of Supply and Services 1981.
Emond, D.P. 'Environmental Law and Policy: A Retrospective Examination of

the Canadian Experience.' In Ivan Bernier and Andree Lajoi, eds. *Consumer Protection, Environmental Law and Corporate Power*. Toronto: University of Toronto Press 1985, 89–181.

Environment Canada Parks Service. *Environment Canada: The Canadian Parks Service Strategic Plan*. Ottawa: Minister of the Environment 1991.

– *State of the Parks 1990 Report*. Ottawa: Minister of Supply and Services 1991.

'Environmental Law Symposium.' *Yale Journal of Regulation* 7/3 (1990).

Fenge, T., and Graham Smith. 'Reforming the Federal Environmental Assessment and Review Process.' *Canadian Public Administration* 12 (1986), 596–605.

Filyk, Gregory, and Ray Cote. 'Pressures from Inside: Advisory Groups and the Canadian Environmental Community.' In Robert Boardman, ed. *Canadian Environmental Policy*. Toronto: Oxford University Press 1992, 60–82.

Foster, Janet. *Working for Wildlife: The Beginning of Preservation in Canada*. Toronto: University of Toronto Press 1978.

Freeman, Myrick A. 'Water Pollution Policy.' In Paul R. Portney, ed. *Public Policies for Environmental Protection*. Washington: Resources for the Future 1990.

Gibson, Robert B. 'Out of Control and Beyond Understanding: Acid Rain as a Political Dilemma.' In Robert Paehlke and Douglas Torgerson, eds. *Managing Leviathan: Environmental Politics and the Administrative State*. Peterborough: Broadview Press 1990, 243–82.

Goodin, Robert E. 'The High Ground Is Green.' *Environmental Politics* 1/1 (Spring 1992), 1–8.

Hampson, Fen Osler. 'Climate Change and Global Warming: The Elusive Search for an International Convention.' Occasional Paper Series, Maxwell School of Public Affairs, Syracuse University, 6 December 1990.

– 'New Wine in an Old Bottle? The International Politics of the Environment.' Paper presented to the Conference on the New World Situation and the Future of Peace, Hobart and William Smith Colleges, Geneva, New York, 27 April 1991.

Hare, Kenneth. 'Environmental Uncertainty: Science and The Greenhouse Effect.' In G. Bruce Doern ed. *The Environmental Imperative*. Toronto: C.D. Howe Institute 1990, 19–34.

Harris, Richard A., and Sydney M. Milkis. *The Politics of Regulatory Change: A Tale of Two Agencies*. New York: Oxford University Press 1989.

Hoberg, George. 'Environmental Policy: Alternative Styles.' In Michael Atkinson, ed. *Governing Canada: Institutions and Policy*. Toronto: Harcourt Brace Jovanovich 1993, 307–42.

– *Pluralism by Design: Environmental Policy and the American Regulatory State*. New York: Praeger 1992.

- 'Representation and Governance in Canadian Environmental Policy.' Paper presented at the Canadian Political Science Association Meetings, June 1991.
- 'Sleeping with an Elephant: The American Influence on Canadian Environmental Regulation.' *Journal of Public Policy* 11 (1991), 107–31.
Hoberg, George, Kathryn Harrison, and Karin Albert. 'The Politics of Canada's Green Plan.' Paper presented at the Canadian Political Science Association Meetings, Ottawa, 7 June 1993.
House of Commons. *Still Waters*. Report of the Sub-Committee on Acid Rain of the Standing Committee on Fiheries and Forestry. Ottawa: Minister of Supply and Services 1981.
- *Time Lost: Demand for Action on Acid Rain*. Report of the Sub-Committee on Acid Rain of the Standing Committee on Fisheries and Forestry. Ottawa: Minister of Supply and Services 1984.
Howatson, Allan. *Business and the Environment: Economic Benefits from Environmental Improvements*. Ottawa: Conference Board of Canada 1991.
Howlett, Michael. 'The Round Table Experience: Representation and Legimacy in Canadian Environmental Policy Making.' *Queen's Quarterly* 97/4 (Winter 1990), 580–601.
Hurrell, Andrew, and Benedict Kingsbury, eds. *The International Politics of the Environment*. Oxford: Clarendon Press 1992.
Ilgen, Thomas. 'Between Europe and America, Ottawa and the Provinces: Regulating Toxic Substances in Canada.' *Canadian Public Policy* 11/3 (1985), 578–90.
Industry, Science and Technology Canada. *Environmental Industries Sector Initiative Workplan for 1989–90*. Ottawa, 1989.
Jacobs, Michael. *The Green Economy*. London: Pluto Press 1991.
Kelley, D.G. 'Discussion Document: Federal-Provincial Committee on Air Quality.' Ottawa: Environment Canada 1990.
Kyba, Patrick. 'International Environmental Relations: Twenty Years of Canadian Involvement.' Unpublished paper, University of Guelph, 1990.
Laplante, Benoît. 'Environmental Regulation: Performance and Design Standards.' In G. Bruce Doern, ed. *Getting It Green: Case Studies in Canadian Environmental Regulation*. Toronto: C.D. Howe Institute 1990, 59–88.
Law Reform Commission. *Policing Pollution*. Ottawa, 1982.
Leiss, William. *The Domination of Nature*. New York: George Braziller 1990.
Lewis, Drew, and William Davis. *Joint Report of the Special Envoys on Acid Rain*. Ottawa: Minister of Supply and Services, January 1986.
Lothian, W.F. *A History of Canada's National Parks*, Vol. 1. Ottawa: Parks Canada 1976.
Lucas, Allistaire R. 'The New Environmental Law.' In Ronald L. Watts and Dou-

glas M. Brown, eds. *Canada: The State of the Federation 1989*. Kingston: Institute of Intergovernmental Relations, Queen's University, 1989, 167–92.

Macdonald, Douglas. *The Politics of Pollution*. Toronto: McClelland and Stewart 1991.

McHallam, Andrew. *The New Authoritarians*. London: Institute for European Defence and Strategic Studies 1991.

MacNeill, Jim, and Robert Munro. 'Environment 1972–92: From the Margin to the Mainstream.' *Ecodecision* 1 (1991), 19–23.

Maslove, Allan, G. Bruce Doern, and Michael Prince. *Federal and Provincial Budgeting*. Toronto: University of Toronto Press 1986.

Mathews, Jessica Tuchman. 'Redefining Security.' *Foreign Affairs* 68/2 (Spring 1989), 162–77.

May, Elizabeth. *Paradise Won: The Struggle for South Morseby*. Toronto: McClelland and Stewart 1990.

Meadows, D.H., D.L. Meadows, J. Randers, and W.W. Behren. *The Limits to Growth: A Report for the Club of Rome's Project on the Predicament of Mankind*. New York: Universe Books 1972.

Morgan, Nancy, Martin Palleson, and A.R. Thompson. *Environmental Impact Assessment and Competitiveness*. Ottawa: National Roundtable on the Environment and the Economy, March 1993.

Munton, Don. 'Dependence and Interdependence in Transboundary Environmental Relations.' *International Journal* 36 (1980–1), 139–84.

Munton, Don, and Geoffrey Castle. 'The Continental Dimension: Canada and the United States.' In Robert Boardman, ed. *Canadian Environmental Policy: Ecosystems, Politics and Process*. Toronto: Oxford University Press 1992.

National Task Force on Environment and Economy. *Report*. Ottawa: Minister of Supply and Services 1987.

Nelles, H.V. *The Politics of Development: Forest, Mines and Hydro-Electric Power in Ontario, 1841–1941*. Toronto: Archer Books 1974.

Nelson, J.G. *An External Perspective on Parks Canada Strategies 1986–2001*. Waterloo: University of Waterloo 1984.

Nemetz, Peter. 'Federal Environmental Regulation in Canada.' *Natural Resources Journal* 26 (1986), 551–608.

Newbery, David M. 'Acid Rain.' *Economic Policy* 5/2 (October 1990), 297–346.

Nitze, William A. *The Greenhouse Effect: Formulating a Convention*. London: Royal Institute for International Affairs 1990.

Olson, Mancur. *The Logic of Collective Action*. Cambridge, MA: Harvard University Press 1971.

Organization for Economic Cooperation and Development. *The OECD Programme on Long Range Transport of Air Pollutants*. Paris, 1977.

Paelhke, Robert. *Environmentalism and the Future of Progressive Politics.* New Haven, CT: Yale University Press 1989.

Paelhke, Robert, and Douglas Torgerson, eds. *Managing Leviathan: Environmental Politics and the Administrative State.* Peterborough: Broadview Press 1990.

Park, Chris. *Acid Rain: Rhetoric and Reality.* New York: Methuen/Routledge 1989.

Pearce, David, A. Markandya, and E. Barbier. *Blueprint for a Green Economy.* London: Earthscan 1989.

Pearce, David, and R.K. Turner. *The Economics of Natural Resources and the Environment.* London: Harvester Wheatsheaf 1990.

Pepper, David. *The Roots of Modern Environmentalism.* London: Croom Helm 1984.

Planning Branch, Treasury Board. 'The Role and Functions of the Department of the Environment and Renewable Resources.' Ottawa, November 1970.

Porter, Gareth, and Janet Welsh Brown. *Global Environmental Politics.* Boulder, CO: Westview Press 1991.

Portney, Paul R., ed. *Public Policies for Environmental Protection.* Washington: Resources for the Future 1990.

Pross, Paul. *Group Politics and Public Policy,* 2d ed. Toronto: Oxford University Press 1991.

Regens, James L., and Robert W. Rycroft. *The Acid Rain Controversy.* Pittsburg: University of Pittsburg Press 1988.

Savoie, Donald. *Regional Economic Development: Canada's Search for Solutions.* Toronto: University of Toronto Press 1986.

Schneider, Stephen H. *Global Warming.* New York: Vintage Books 1990.

Science Council of Canada. *Canada as a Conserver Society.* Ottawa, 1977.

– *Policies and Poisons.* Ottawa, 1977.

Schrecker, T.F. 'The Canadian Environmental Assessment Act: Tremulous Step Forward or Retreat into Smoke and Mirrors.' *Canadian Environmental Law Reports* 5 (March 1991), 192.

– 'Of Invisible Beasts and the Public Interest: Environmental Cases and the Judicial System.' In Robert Boardman, ed. *Canadian Environmental Policy: Ecosystems, Politics and Process.* Toronto: Oxford University Press 1992, 83–105.

– *Political Economy of Environmental Hazards.* Ottawa: Law Reform Commission of Canada 1984.

– 'Resisting Regulation: Environmental Policy and Corporate Power.' *Alternatives* 14/1 (1987), 9–21.

Sims, Andrew. 'If Not Then, When? Non-Governmental Organizations and the Earth Summit Process.' *Environmental Politics* 2/1 (Spring 1993), 94–101.

Sinclair, William F. 'Controlling Effluent Discharges from Canadian Pulp and Paper Manufacturers.' *Canadian Public Policy* 16/1 (March 1991), 86–105.

– *Controlling Pollution from Canadian Pulp and Paper Manufacturers: A Federal Perspective*. Ottawa: Environment Canada 1988.

Skogstad, Grace, and Paul Kopas. 'Environmental Policy in a Federal System: Ottawa and the Provinces.' In Robert Boardman, ed. *Canadian Environmental Policy: Ecosystems, Politics and Process*. Toronto: Oxford University Press 1992, 43–59.

Smith, Douglas A. 'Defining the Agenda for Environmental Protection.' In Katherine A. Graham, ed. *How Ottawa Spends: 1990–91*. Ottawa: Carleton University Press 1990, 113–36.

Swaigen, John, and Gail Bunt. *Sentencing in Environmental Cases*. Ottawa: Law Reform Commission of Canada 1985.

Swift, Jamie. *Cut and Run: The Assault on Canada's Forest*. Toronto: Between the Lines 1983.

Task Force on Program Review. *Environmental Quality Strategic Review: A Follow-on Report*. Ottawa: Minister of Supply and Service 1986.

– *Improved Program Review: Environment*. Ottawa: Minister of Supply and Services 1986.

– *Programs of the Minister of the Environment*. Ottawa: Minister of Supply and Services 1986.

Taylor, S. *Making Bureaucracies Think: The Environmental Impact Statement Strategy of Administrative Reform*. Stanford, CA: Stanford University Press 1984.

Thomas, Caroline. 'The United Nations Conference on Environment and Development of 1992 in Context.' *Environmental Politics* 1/4 (Winter 1992), 250–61.

Thompson, Andrew. *Environmental Regulation in Canada: An Assessment of the Regulatory Process*. Vancouver: Westwater Research Centre 1980.

Toner, Glen. 'The Canadian Environmental Movement: A Conceptual Map.' Unpublished paper, Carleton University, 1991.

– 'Stardust: The Tory Energy Program.' In Michael J. Prince, ed. *How Ottawa Spends: 1986–87*. Toronto: Methuen 1986, 119–48.

– 'Whence and Whither: ENGOs, Business and the Environment.' Carleton University Centre for the Study of Business–Government–NGO Relations, Working Paper 1, 1991.

Victor, Peter A. *Environmental Protection Regulation: Water Pollution and the Pulp and Paper Industry*. Ottawa: Economic Council of Canada 1981.

Vogel, David. *National Styles of Regulation: Environmental Policy in Great Britain and the United States*. Ithaca, NY: Cornell University Press 1986.

Weale, Albert. *The New Politics of Pollution*. Manchester: Manchester University Press 1992.

Webb, Kernaghan. 'Between Rocks and Hard Places: Bureaucrats, the Law and Pollution Control.' *Alternatives* 14 (May/June 1987), 4–14.

– *Pollution Control in Canada: The Regulatory Approach in the 1980s*. Ottawa: Law Reform Commission of Canada 1986.

Whalley, John. 'The Interface between Environmental and Trade Policies.' *The Economic Journal* 101 (March 1991), 180–9.

Whittington, Michael. 'The Department of the Environment,' In G. Bruce Doern, ed. *Spending Tax Dollars*. Ottawa: School of Public Administration, Carleton University, 1980, ch. 7.

Wilks, Stephen. 'Administrative Culture and Policy Making in the Department of the Environment.' *Public Policy and Administration* 2/1 (Spring 1987), 25–41.

Wilson, Jeremy. 'Green Lobbies: Pressure Groups and Environmental Policy.' In Robert Boardman, ed. *Canadian Environmental Policy: Ecosystems, Politics and Process*. Toronto: Oxford University Press 1992, 109–25.

World Commission on Environment and Development. *Our Common Future*. Oxford: Oxford University Press 1987.

ENVIRONMENT CANADA DOCUMENTS AND PAPERS

'Amendments to the Environmental Assessment and Review Process: Submission to the Interdepartmental Committee on the Environment.' 14 January 1976.

'A Ten Year Action Plan for Environment Canada, 1975–1985.' February 1974.

'Briefing Note on the Canadian Coalition on Acid Rain.' July 1982.

'Briefing Notes for the Deputy Minister: Indian and Northern Affairs.' 4 July 1975.

'Briefing Notes for the Deputy Minister: Ministry of Transport.' 4 July 1975.

'Briefing Notes for the Minister on Memorandum to Cabinet on EARP.' 12 November 1976.

Canadian Environmental Protection Act Report of the Period Ending March 1990. Ottawa: Minister of Supply and Services 1990. (Also reports for years ending March 1991 and March 1992).

'Canadian Position on the North American Waterfowl Management Plan.' Letter sent by Charles Caccia, Minister of the Environment, to the provincial Environment Ministers, 15 March 1984.

Departmental Briefing Book, Vol. 1 (Departmental Overview), March 1987.

'Departmental Environmental Assessment Policy Review.' FEARO, 2 May 1991.

'Department of the Environment: Preliminary Statement of Goals for 1980.' 14 January 1972.

'Draft Strategic Plan.' 6 April 1981.

'Draft White Paper on the Environment,' Draft No. 1, 12 October 1978.

'EARP Review – Status Report.' 30 June 1983.

'EMR Review of: "A Study of Environmental Concerns: Offshore Oil and Gas Drilling and Production.' 12 February 1979.

'Environmental Assessment Panel,' sent by Reed Logie to the Deputy Minister, 25 April 1974.

'Environmental Evaluation Process,' memorandum sent by Gislain M. Gauthier, Quebec Region, to Deputy Minister Blair Seaborn, 2 October 1975.

Environment Canada Chronologies: 1970–1990. April 1991.

'Environment Canada's Perspective on the Management of Wildlife in Canada: A Discussion Paper on Issues and Initiatives.' 25 April 1985.

'Environmental Protection Service Analysis of Departmental Action Plan.' 15 January 1973.

'Federal Environmental Responsibilities: Mandate of the Minister of the Environment.' 15 August 1978.

'Federal-Provincial Relations within Environment Canada: An Overview.' Intergovernmental Affairs Directorate, March 1982.

'Financial Situation of the Department,' memorandum sent by the Deputy Minister to the Minister of the Environment. 28 June 1979.

'Financial Situation of the department,' memorandum sent by the Deputy Minister to the Minister of the Environment. 13 August 1979.

'Highlights of Successive Changes (Reductions) in the Office of the Science Advisor.' 1979.

'Impact of Resource Reductions Imposed Since 1975/76 on the Corporate Policy Group and the Office of the Science Advisor.' 1979.

'Letter from Deputy Minister of Regional Industrial Expansion to Deputy Minister of Environment.' 19 August 1986.

'Letter from Prime Minister Trudeau to Len Marchand, Minister of the Environment, regarding the separation of Fisheries from Environment.' 9 November 1978.

'Letter from R.G. Skinner, Coordinator, Office of Environmental Affairs, EMR, to Jag Maini, Chairman, Energy Review Group, DOE.' 27 November 1979.

'Major Priority Environmental Issues for the Eighties.' 7 December 1979.

'Media Trends.' July 1979.

'Memorandum from Assistant Deputy Minister to Deputy Minister: Situation Report AES 1980/81.' 17 August 1981.

'Memorandum from Assistant Deputy Minister Parks Canada to the Deputy Minister of Environment Canada.' 28 March 1985.

'Ministerial Briefing Book: Norther American Waterfowl Management Plan – Evolution and Rationale of Funding Proposal to the Western Diversification Office.' 3 March 1988.

'Ministerial Briefing Paper on CEPA Legislation.' 5 February 1986.

'Memorandum from Deputy Minister Robert Shaw to Director, Personnel
  Branch, on the Organization of DOE.' 26 March 1971.
'Notes for Remarks by the Honourable Tom MacMillan, P.C., M.P., Federal Min-
  ister of the Environment, to the Legislative Committee on Bill C-74, The Cana-
  dian Environmental Protection Act.' 25 November 1987.
Office of Science Advisor, *Maintenance and Utilization of Science in the Department
  of the Environment*, September 1983.
'Outline of an Action Plan for Environment Canada.' 12 December 1972.
'Outline of an Action Plan for Environment Canada: A Review by the Fisheries
  Service.' 16 January 1973.
'Ownership of Wildlife,' letter sent to Mr Gorden, ADM, Alberta Department of
  Recreation, Parks and Wildlife, by Mr H. Boyd, Director Migratory Birds
  Branch, CWS, 23 April 1979.
Policy Respecting Science and Technology, 20 October 1984.
'Preliminary Report, Inhouse Operations: Department of the Environment.'
  24 June 1974.
'Press Release: Billion Dollar Canada–U.S. Waterfowl Management Plan
  Released for Public Comment.' 2 December 1985.
'Problems Facing Wildlife Habitate Management on Canadian Forest Lands.'
  February, 1981.
'Program Review,' memorandum sent by the Deputy Minister to the Minister of
  the Environment, 16 August 1979.
'Proposal on the Environmental Assessment Panel,' sent by Jack Davis to the
  Treasury Board, 10 January 1974.
'Proposed Cabinet Memorandum Recommending Amendments to the Environ-
  mental Assessment and Review Process.' 12 November 1976.
'Protection of the Federal Fisheries from Pollution.' 9 July 1982.
'Review of the Environmental Assessment and Review Process,' memorandum
  sent by Andre A. Grignon, Director, Federal Provincial Relations Office, to
  Reed Logie, Chairman of FEARO, 7 April 1975.
'Second Draft Statement on Environmental Contaminants Legislation,' Cross
  Mission Task Force on Environmental Contaminants, 22 June 1972.
'Senior Assistant Deputy Ministers Review of the Environmental Management
  Services Activities,' 1974.
'Statement by Dr James H. Patterson on the North American Waterfowl Manage-
  ment Plan.' 10 September 1987.
'1984–1989 Strategic Plan.' October 1983.
'Submission Guidelines: Habitat Conservation, Restoration and Enhancement,'
  Wildlife Habitat Canada, January 1988.
'Summary Listing of the Outline of the Action Plan.' 1972.

'The Department of the Environment,' Lands Directorate, 1970.

'The Montebello Goals.' April 1971.

'The North American Waterfowl Management Plan.' 21 March 1988.

'Waterfowl Management and Crop Depredation.' 29 April 1980.

'Waterfowl Management Plan for Canada: An Overview.' March 1983.

'Waterfowl Management Plan for Canada,' Proposal sent by James Patterson of CWS to A.G. Loughrey, Director General of CWS, 5 March 1981.

'Whither We Goeth?' 1975.

'Working Paper on Proposed Environmental Contaminants Act for Inter-departmental Discussion and Discussion with Provincial Authorities.' 18 April 1973.

Index

acid rain 7, 13, 43, 44, 55, 58, 63, 64, 75, 85, 91, 92, 107, 110, 112, 114, 117, 128, 129, 130, 131, 134, 135, 136, 137, 139, 148, 245, 246; agreements on 153–4; and Air Pollution Control Directorate 149, 152; and Atmospheric Environment Service 35–6, 149, 151–2, 153, 163, 167; bilateral stalemate on (1982–9) 155–8; causes of 155; as DOE priority 148–9, 152–3, 154, 159, 161–2, 165, 167, 234; domestic breakthrough on 158–63, 166; early focus on 149, 151–5, 166; key events in policy 149, *150–1*; and media process 156, 231–2, 234; and PMO funding 162; public view of 148; and Quebec programs 161, 163; U.S. action on 163–6; *see also* federal-provincial relations
Acid Rain Abatement Policy 163
Acid Rain Coalition. *See* Coalition on Acid Rain
Agriculture Canada 20, 22; transfer of Forestry Service to 29
Air Pollution Control Association 153

Air Pollution Control Directorate 89, 90, 91, 152–3
air quality (urban), DOE action on 12
Alberta, response of to Green Plan 98–9; *see also* federal-provincial relations; Oldman River project (Alberta)
Algoma 155, 161
Anthony, Russ 19
Arctic: clean-up 79; sovereignty 36
Arctic Waters Pollution Prevention Act 67
asbestos industry, political clout of 108
Atikokan (Ontario), power plant construction 153
Atlantic Canada. *See* federal-provincial relations
Atmospheric Environment Service 16, 17, 89; budget of 35, 36; history of 33; integration of into other areas 29, 34; and international relations 140; organizational components of 34, 35, 36; role of 33–6, 37, 127, 133; *see also* acid rain
Atmospheric Research Directorate 34, 151–2

Atomic Energy Control Board 70–1
Atomic Energy of Canada Ltd 70
Austria 137–8
Axworthy, Lloyd 19

Barrett, Dave 178
Bateman, Robert 175, 176
Bates, David 152
Bell Canada 115
Bennett government (BC) 179
Berger, Justice Thomas 62
Berger Commission of Inquiry into
  the Alaska Pipeline 8, 62, 69, 198
Bhopal chemical disaster (India) 14,
  115, 221
Bill C–13 209; see also Canadian Envi-
  ronmental Assessment Act
Bill C–78 209; see also Canadian Envi-
  ronmental Assessment Act
Bird, Peter 133
Blais-Grenier, Suzanne: tenure of 13,
  46–7, 49, 55, 63–4, 65; and South
  Moresby decision 173, 176, 177,
  181, 182, 234; and Wildlife Service
  budget cuts 33, 47, 77, 143
Bouchard, Lucien: appointment of as
  environment minister 14, 74, 234;
  and environmental assessment
  207–8; and Green Plan 57, 98, 119;
  resignation of 120; tenure of 49–50,
  56
Boundary Waters Treaty (Canada–
  U.S.) 127
British Columbia: Burrard Inlet
  bridge decision 195, 237; see also
  federal-provincial relations; Okanagan Valley; Skagit Valley project;
  South Moresby Park
Brooks, David 19, 55
Bruce, Jim 34–5, 88, 128, 135, 136

Bruce Peninsula (Ontario), and proposed national park 187
Brundtland Commission (World
  Commission on Environment and
  Development) 5, 45, 48, 58, 64, 114,
  124, 141, 142–3, 147, 207, 232, 236,
  243
Brydon, Jim 133
Bush, George 164
Bush Administration 243
business interests: and DOE relationship 100, 107–10, 111–16 (closing of
  gap between), 118–22 (Green Plan),
  137, 209 (environmental assessment); and ENGOs 111–16, 137,
  143; and Rio Summit 143; see also
  Department of the Environment;
  Environmental Protection Service

Caccia, Charles 45–6, 112–13, 135,
  136, 204, 206, 220, 234; and South
  Moresby 173, 176, 180
Camp, Dalton 185
Campbell, Kim 51
Canada Centre for Inland Waters 30,
  88
Canada Land Inventory 32
Canada–U.S. Great Lakes Water Quality Agreement 216, 243
Canada–U.S. relations 31, 124, 125–6,
  129–32; and acid-rain issue 148–67;
  and agricultural run-off 130; and
  Great Lakes issue 155, 213, 216,
  243; and pollution treaties
  between 127; and pulp and paper
  industry 130; see also Boundary
  Waters Treaty (Canada–U.S.); Commission for Environmental Cooperation; environmental non-government organizations; Garrison

Diversion project (Manitoba); International Joint Commission; North American Commission on the Environment; North American Free Trade Agreement; Shamrock Summit
Canada Water Act 21, 22, 30, 31, 87, 89
Canada Wildlife Act 33
Canadian Assembly on National Parks and Protected Areas 182
Canadian Chemical Producers Association 115
Canadian Council of Ministers of the Environment 91–2, 114, 139
Canadian Council of Resource (and Environment Ministers) 91–2
Canadian Environmental Advisory Council 76, 105–6, 207, 227
Canadian Environmental Assessment Act 80, 98
Canadian Environmental Network 104, 106, 114, 181
Canadian Environmental Protection Act (CEPA) 7, 11, 49, 58, 85, 87, 92, 97, 99, 115, 245, 246–8; acts superseded by 21; components of 224; DOE goals in 223; and federal–provincial agreements under 94, 225–8; and Fisheries Act 86–7, 242; and industry monitoring 115; passage of 14, 109, 222, 228; purpose of 21–2; and Regulatory Impact Assessment Study requirement 227–8
Canadian Forestry Service. See Forestry Service
Canadian Meteorological Service 151
Canadian Nature Federation 102–3, 106, 116, 174

Canadian Parks and Wilderness Society 174, 187
Canadian Pulp and Paper Association 115
Canadian Wildlife Federation 102, 174, 207
Canadian Wildlife Service 67; see also Land, Forest and Wildlife Service
carbon tax 122
Carson, Rachel 29, 103
Carter Administration 154, 156
CFC agreements, negotiations on 125, 126, 137; see also ozone layer
Chant, Donald 19
Charest, Jean 51
Charter of Rights and Freedoms 116
chemical industry, improved record of 115
Chernobyl nuclear-reactor meltdown 14, 36, 118
Chrétien, Jean 62, 170
Clark, Joe 63, 176, 179
Clean Air Act (Canada) 21, 22, 89, 90, 154, 213–14
Clean Air Act (U.S.) 136, 139, 149, 152, 154, 155, 164
Clean Water Act 213
Clinton Administration 125, 126, 143–7, 243
Club of Rome 19–20, 102
Coalition on Acid Rain 44, 107, 112, 154, 165
Collinson, Jim 172, 184, 185
Commission for Environmental Cooperation 146
computer sector 115
constitution, and reference to environment 83, 99
Convention on Long-Range Transboundary Air Pollution 153–4

Corporate Policy Branch, negotiations
by of protocols 137
Countdown Acid Rain program
(Ontario) 159
Cullen, Bud 208

Darling, Stan 166
Davidson, Al 19, 27, 78, 170, 174
Davis, Bill 151, 163–4
Davis, Jack 38, 39–41, 193, 195, 213,
214–15, 217, 234, 237
de Cotret, Robert 14, 50–1, 120
Department of the Environment
(DOE): budget of 26, 27, 47, 48,
64–6, 81, 93, 76, 77, 107, 170–1, 172;
and business interest groups 100,
107–10, 111–16, 122–3, 137, 143;
compared with U.S. EPA 15; creation of 3, 16–20, 52–3; and DEA
63, 124, 126, 128, 133–4, 137; and
Department of Finance 50, 122; and
DIAND 62, 66–8, 69, 81–2; and
Fisheries components/depart-
ments 13, 16, 17, 22–6, 28, 29, 31, 37,
40, 42, 44, 63, 95, 168–9, 178, 200;
employees of (background) 74, 75;
and EMR 20, 30, 31, 66, 68–71, 81–2;
and ENGOs 100–7, 111–16, 122–3,
137, 143, 242–3; goals and priorities
of 52, 53–4, 55, 56, 151, 152–3; and
internal conflicts 233–4; and international agreements 137–41; international role of 124–47; and ISTC
66, 71–4, 81–2; legislative base of
17, 18, 20–2, 26; literature on 9–10;
mandate of 7, 17–20, 60, 168; media
coverage of 117–18, 156; and
national parks 171, 183–4, 188, 189;
organizational resources of 6–7;
organizational structure of 16,
22–3, 37; and PCO 62–3, 64; and
PMO 17, 62–4, 124; political leadership of 13; and public opinion 100–
1, 111–12, 116–18, 122, 123; and reg-
ulatory-responsibility position
220–1; role of in Brundtland, Rio,
Stockholm, World Conservation
Strategy 147; science functions of
74–8; State of Environment reports
of 113; summarized 12–15, 58–9,
61, 241–3; and sustainable development concept 143; and 30 Per Cent
Club leadership role 136–7; see also
acid rain; air quality (urban); Atmospheric Environment Service;
Canadian Environmental Protection Act; Environment Services;
Environmental Assessment and
Review Process; Environmental
Protection Service; Federal Environmental Assessment and Review
Office; federal-provincial relations;
Finance and Administration Service; Fisheries Service (Fisheries
and Marine Service); Forestry Service; Great Lakes; Inland Waters
Directorate; Land, Forest and Wildlife Service; Lands Directorate;
Parks Service/Parks Canada; phosphates; Policy, Planning and
Research Service; pollution (air,
water); South Moresby Park; toxic
contaminants; Water Management
Service; Wildlife Service
Department of Energy, Mines and
Resources 20, 30, 31, 66, 68–71,
81–2; involvement of in acid-rain
study 161–2
Department of External Affairs 71;
and acid-rain issue 156; and DOE

63, 124, 126, 128, 133–4, 137; lobby-
ing of by energy industry 139; and
Rio Summit 81
Department of Finance 50, 122
Department of Fisheries (and Fisher-
ies components) 13, 16, 17, 22–6,
28, 29, 31, 37, 40, 42, 44, 63, 95,
168–9, 178, 200
Department of Health and Welfare
31, 89
Department of Heritage 23, 169
Department of Indian Affairs and
Northern Development 20, 22–3,
62, 66–8, 69, 81–2, 169, 170; and
Green Plan 81
Department of Industry, Science and
Technology (ISTC) 66, 71–4, 81–2
Department of Industry, Trade and
Commerce 20, 71, 72
Department of Regional Economic
Expansion 71, 72
Department of Regional Industrial
Expansion 71, 72–3, 161–2
dioxins, and the Great Lakes 220
Dome Petroleum 35
Dominion Forest Preserves and Parks
Act 169
Dow Chemical 116, 221
Ducks Unlimited Canada 102–3

Earth Summit. See Rio
ecologism, philosophy of 103–4
Economic Commission for Europe
132, 134, 136, 139, 154
ecosystems, defined 8–9
Ellesmere Island National Park
Reserve 187
emissions. See NOx and VOCs
emissions control, early 212–16
energy crises 68–9, 142, 198

energy industry 108, 122, 139
Energy Options 70, 114
Energy Probe 103
energy tax 98
Environment Services, creation of
25–6
environment taxes 122
Environmental Assessment and
Review Process: barriers to
197–200; and Brundtland Commission 207; and Cabinet 190, 190–1
(1977), 194–5 (1973); described 190,
191; early developments in 190,
192–7; and ENGOs' pressure 203;
and federal-provincial relations
206; as guidelines order 205–6; legislation of 204, 208–9; legislative
base for 190, 193; and Macdonald
Commission recommendation 207;
and provincial processes 199–200;
and public involvement 197–8,
203–4, 205–6; and reforms 197–202
(1977), 206–8; see also Federal Environmental Assessment and Review
Office
environmental bill of rights 120
environmental concerns, international. See Brundtland Commission;
Canada–U.S.relations; Department
of External Affairs; General Agreement on Tariffs and Trade; Great
Lakes; North American Free Trade
Agreement; Rio Earth Summit; SO2
30 Per Cent Club; Stockholm Conference; World Conservation Strategy
Environmental Contaminants Act 13,
21–2, 32, 109, 216–18, 220, 221, 228
Environmental Management Service
87–8

environmental non-government organizations (ENGOs) 43, 47, 48–9,
235, 238; and business 111–16, 137,
143; and data collection 131; and
DOE relationship 100–7, 111–16,
122–3, 137, 143, 242–3; education
activities of 101–2; and environmental assessment 203, 209, 210;
funding of 106–7, 111, 112; and
Green Plan 101, 118–22, 233, 243;
growth of 104; history of 103; and
industry 108; and international
issues 130; lobbying activities of
101–2; and media 102; and NOx-
VOCs process 140; portrayal of
United States by 131; and public
consultation 111–12; and Rio Summit 143; and South Moresby Park
174, 181, 182–3, 185; *see also* Canadian Nature Federation; Pollution
Probe
Environmental Protection Act 48
Environmental Protection Agency
(U.S.) 156; and bilateral consultation on acid rain 153; compared
with DOE 15, 130, 242
Environmental Protection Service 16,
17, 22, 26, 75, 86; and acid rain 157;
and business-interest groups
107–8, 110, 123; cooperation of with
Inland Waters 31; and Environmental Contaminants Act 218;
early focus of 212–16; and federal-
provincial accords 96, 219; formation of 31–2; and friction with Forestry Service 28–9; and IJC role
127; and industry regulation 108–9;
leadership of 87; linked with Environmental Conservation Service
29; organization of 30, 31–2, 216;

policy at 111; regulatory role of
212–16, 226–7, 228–9; *see also* Canadian Environmental Protection Act
Environmental Quality Act (Quebec)
199
Environmental Quality Directorate,
establishment of 16
Environmental Quality Policy Framework 48
Exxon Valdez oil spill 14

Falconbridge 155, 161, 162
Federal Environmental Assessment
and Review Office 128, 191, 196–7,
198–9, 205, 209, 210
Federal-Provincial Accords for the
Protection and Enhancement of
Environmental Quality 95
Federal-Provincial Advisory Committee on Air Quality 89, 90, 91
federal-provincial relations 6, 7–8, 13,
22, 30–1, 45, 71, 83–5, 89–94, 95–7,
97–9; 130; and acid rain 148–9,
150–1, 154, 158–63; and air issues
214; and environmental assessment 209–10; and environmental
accords 95, 96; and Environmental
Protection Act 217; on Western
resources 169. By province/
region: Alberta 69, 89, 93, 97;
Atlantic Canada 86–7, 93; BC 86–7,
93, 97, 186; Manitoba 89; Ontario
88, 93, 97; Quebec 86–7, 93, 97;
Saskatchewan 89, 97; *see also* constitution; Environmental Assessment and Review Process;
Environmental Protection Service;
Green Plan; pollution (air, water)
Finance and Administration Service
16

Fisheries Act 13, 21, 22, 23, 25, 26, 32, 86–7, 93, 97, 105, 109, 178, 182, 208, 213, 214, 218–21 (reform)

Fisheries and Marine Service 42; creation of 25–6; departure of from DOE 28

Fisheries Service 13, 16, 17, 42, 44, 63, 95, 200, 213; and Environmental Protection Service 31–2; and friction within DOE 23–5, 37; and Parks Service, transfer with 22–3, 26, 29, 168–9; policy objectives of 23–5

Forestry Service 17, 72; and friction with Environmental Protection Service 28–9, 37; and friction within DOE 28; transfer of to Agriculture Canada 29

Fraser, John 19, 43, 47, 64, 106, 107, 112, 153, 154, 165, 176, 177, 179–80, 182, 202, 207, 220

Friends of the Earth 103

Friends of the Oldman River Society 208

Fulton, Jim 176, 179–80

Garrison Diversion project (Manitoba) 129, 130–1, 135

Gauthier, Ghislain M. 196

General Agreement on Tariffs and Trade 125, 144–6, 147

Geological Survey of Canada 68

Gerin, Jacques 46, 48, 78, 96, 112

Germany 135, 137–8

Giles, Walter 88

global warming 7, 8, 14, 98, 126, 140, 164, 246

Good, Len 49, 50, 56

Goods and Services Tax 97–8, 122

Government Organization Act 200

Great Lakes water quality 8, 31, 85, 88, 155; DOE action on 12, 75, 110, 151

Great Lakes Water Quality Agreement 125, 126–9, 130, 132, 232

Great Lakes Water Quality Board 153

Green Parties 137–8, 140

Green Party (West Germany) 135, 137–8

Green Plan (1990) 8, 13, 38, 44, 244; Alberta's response to 98–9; and business interests 110, 121–2; content of 78–80, 119, 120; and ENGOs 101, 118–22, 233, 243; and federal-provincial relations 97–9; framework of 57–8; funding of 14, 57, 64, 78–82, 97, 98, 99, 119; future place of 239–40; launch of 3, 14, 51, 164, 233; media response to 120; and Priority Substances List 227; process of 48, 50–1, 57, 58, 59, 69, 73, 74, 118, 123; and public role in 101, 118–20; targets of 80

green taxes 120

Greenpeace Canada 103

Greenprint for Canada 119

Grignon, André A. 197

G-7 Toronto Economic Summit meetings 5, 140, 236–7

Hagersville (Ontario) tire fire 14

Haida Nation: and logging blockade 182–3; park management role of 186; traditions of 175, 189

Hartle, Douglas 18

Helsinki Protocol 136

Historic Sites and Monuments Board 169

Hollins, John 133

Hurley, Adele 150, 165

Hurricane Hazel 88
Illinois 128
Imperial Oil 111
INCO 110, 155, 160, 161, 162
Indian and Eskimo Affairs (DIAND
    program) 67
industries, clean 115
Industry, Science and Technology
    Canada 66, 71–4, 81–2
Inland Waters Directorate 86, 87–8,
    127, 128, 129, 216; cooperation of
    with Environmental Protection Ser-
    vice 31; formation of 30–1
Inquiry on Federal Water Policy 221
International Convention for the Pre-
    vention of Pollution from Ships 133
International Joint Commission 30,
    126, 127–9, 130, 153, 213
International Maritime Organization
    133
International River Improvements
    Act 208
International Trade Canada, and Rio
    Summit 81
International Union for the Conserva-
    tion of Nature 133, 141, 236

James Bay Hydro projects 97

Kierans, Tom 70

Lake St Clair 213
Land, Forest and Wildlife Service 16,
    47
Lands Directorate: and DOE creation
    17, 19; as EARP vehicle 195–6; inte-
    gration of with Wildlife Service 32
Laurier, Wilfrid 102
lawyers, environmental 17, 19
Layt, George 55

LeBlanc, Romeo 42, 46, 63, 87, 153,
    199, 219–20
Lewis, Drew 151, 163–4
Limits to Growth (Club of Rome) 20, 102
lobby, environmental 47; see also
    environmental non-government
    organizations; business interests
Logie, Reed 196
long-range transport of airborne pol-
    lutants (LRTAP) program 35, 92,
    149, 152, 153, 166; see also acid rain
Love Canal (New York State) 220
Lucas, Ken 25, 42, 87

McBean, Gordon 153
McCrory, Colleen 180
Macdonald Royal Commission 143,
    207
MacMillan, Charles 182
MacMillan, Tom 47–8, 49–50, 56, 111,
    112, 114, 186, 220–2, 223; and South
    Moresby decision 168–9, 173, 175,
    176, 177, 179–80, 182, 183, 184, 185,
    207, 234
MacNeill, Jim 124, 134–5, 142
Maine 164
Manitoba 129; see also federal-provin-
    cial relations; Garrison Diversion
    project
Manson, Alex 157
Marchand, Jean 41–2
Marchand, Leonard 43, 201
May, Elizabeth 111, 112, 177, 178, 183,
    184, 185
Mazankowski, Don 185
media coverage, process: and acid
    rain 156, 231–2, 234; of Clean Air
    Act 213–14; of DOE 117–18, 156;
    and ENGOs 102; of environmental
    issues 117; of Green Plan 120; of

NOx-VOCs process 140; of South
  Moresby 231–2
Mexico, and NAFTA provisions 146
Michigan 128
Migratory Birds Convention Act 33
mining sector: and Environmental
  Protection Service 123; industry
  regulation of 8
Ministers of the Environment 38–9;
  see also Blais-Grenier; Bouchard;
  Caccia; Davis; Charest; de Cotret;
  Fraser; LeBlanc; MacMillan; Mar-
  chand, Jean; Marchand, Leonard;
  Roberts; Sauvé
Ministry of State for Economic and
  Regional Development 71
Minnesota 153
Mississauga railway derailment 220
Mitchell, George 164
Montreal Protocol on Substances that
  Deplete the Ozone Layer 8, 125,
  139–40, 144
Morley, Greg 19
Mulroney, Brian 46, 49–50, 56, 63, 73,
  119, 159, 162, 163; and South
  Moresby: 175, 176, 178, 182, 185, 234
Munro, David 124, 133, 142

NAFTA. See North American Free
  Trade Agreement
National Energy Board 55
National Energy Program (1980) 44,
  69–70, 98, 122, 161
National Environmental Policy Act
  (U.S) 192–3
national historic parks and sites, his-
  tory of 169
National Parks Act 169, 187
National Task Force on Environment
  and Economy 114, 143

Native organizations: and Green
  Plan 119
Native Peoples, interests of 43, 62, 67,
  81, 123, 172, 175, 182–3, 186, 189; see
  also Haida Nation
Nature Conservancy of Canada
  102–3
Navigable Water Protection Act 208
Nelson, Gordon 186
New Brunswick. See federal-provin-
  cial relations (Atlantic)
New Directions Group 116
New England states 155
New York State 128, 155, 160, 220
Newfoundland 213; and federal-pro-
  vincial accords 95, 96; see also fed-
  eral-provincial relations (Atlantic)
Niagara Institute 112–13, 114, 162
Nielsen Task Force 47–8, 55, 56, 63,
  172, 180
nitrogen oxides. See NOx and VOCs
Nixon, Buzz 18
Nixon Administration 156, 192
Noranda 116, 162
North American Commission on the
  Environment 145
North American Free Trade Agree-
  ment 125, 126, 143–7; key provi-
  sions of 145
North Dakota 129
Northern Affairs (DIAND program)
  67–8
Northern Ontario, PCB spill in 221
Norton, Keith 160, 61
Norway 152, and 30 Per Cent Club
  136
Nova Scotia, and acid-rain programs
  163; see also federal-provincial rela-
  tions (Atlantic)
NOx (oxides of nitrogen) and VOCs

(volatile organic compounds) 114, 125, 246; international agreements on 137–40; negotiations on 126

nuclear issues 70–1, 118

Ocean Dumping Control Act 218
Ohio 128, 155, 213
Ohio Valley 158
Okanagan Valley (BC) 85
Oldman River project (Alberta) 94, 97, 116, 208–9
Ontario: and acid rain 154, 158, 159–61, 163; mercury discharges in 95; see also Atikokan; Bruce Peninsula; federal-provincial relations; Great Lakes; Hagersville; Lake St Clair; Mississauga; Northern Ontario; St Clair River
Ontario Hydro 110, 152, 155, 161
Organization for Economic Cooperation and Development 132, 134–5, 235; and acid rain 152; Environment Committee of 142–3
Osbaldeston, Gorden 194
ozone layer 7, 8, 90, 126, 140

Pacific Rim National Park, 183, 184
Parks Service (formerly Parks Canada) 17, 246; budget of 55, 170–1, 172; as component of DOE 13, 37, 63, 67, 188; and DOE consultation policy 111; and DOE mandate 233; DOE takeover of 169–71; and Fisheries swap 22–3, 26, 168–9; and friction within DOE 26–8; goals and mandate of 171–4, 187; and interest groups 110–11, 123; legislative history of 168–71; name change of 29; policy of 7, 175; review of policies,

plans 186–8; see also South Moresby Park
Parks 2000 187
Parliament, powers of (environmental affairs) 84
Pearse Commission 45, 221
Pearson, Lester 125
Pelton, Austin 175, 177, 181, 182–3
Pennsylvania 155
Pépin, Jean-Luc 46
Perley, Michael 150, 165
pesticides 97
Peterson, David 159
Petro-Canada 35
phosphorous pollutants (phosphates), DOE action on 12, 88, 110
policy sector, defined 10–11
Policy, Planning and Research Service 16
pollution (air): Canada–U.S. treaty on 127; DOE action on 12; and federal-provincial relations 89–91, 92, 99; and regulatory issues 109; see also acid rain; NOx and VOCs
Pollution Probe 103, 116, 123
pollution (water): Canada–U.S. treaty on 127; DOE action on 12, 24, 25, 26; and federal-provincial relations 85–9, 99; and Fisheries Act 23, 26; and regulatory issues 109; see also acid rain; Great Lakes
preservationism, philosophy of 103–4
Prince, Alan 128
Prince Edward Island. See federal-provincial relations (Atlantic)
Protocol on Oxides of Nitrogen 125
public involvement, opinion: on acid rain 148; and DOE 100–1, 111–12, 116–18, 122, 123; and ENGOs 111–12; on Environmental Assess-

ment and Review Process 197–8,
203–4, 205–6; on environmental
issues 116–18; on environmental
protection 222–3; on Green Plan
101, 118–20
pulp and paper sector: and Environmental Protection Service 123;
improved record of 115; and industry regulation 8; lobbying power of
108; modernization grants to industry by 72; and recession (1981–2)
13; and U.S. relations 130

Quebec: environmental assessment
in 199; *see also* acid rain; federal-
provincial relations; Environment
Quality Act; James Bay; St Basile-le-
Grand
Queen Charlotte Islands. *See* South
Moresby Park

Rafferty–Alameda dams
(Saskatchewan) 93–4, 97, 116; court
case over project 208–9
Rayonier Canada (B.C.) Ltd 175,
178–9
Reagan Administration 131–2, 135,
145, 146, 148, 154, 155–8, 163, 167,
243
recession (1981–2) 13, 142, 156, 202
regulation, environmental 107–10; *see
also* Department of the Environment; Environmental Protection
Service; pollution (air; water); toxic
substances
Regulatory Impact Assessment
Study 228
Rejhon, George 166
Richardson, Miles 175
Rio Earth Summit (1992) 3, 5, 15, 51,

58, 81, 124, 140, 141, 143, 147, 232,
236, 241, 243, 244
Roberts, John 43–4, 45, 112, 160, 202,
204
Robinson, Raymond 128, 157, 202–4,
205
Rogers, Stephen 177, 183
Roots, Fred 76, 133
Royal Commission on Economic
Union and Development Prospects
for Canada. *See* Macdonald Commission
Royal Commission on Fish and Game
102
Royal Society of Canada, review by of
bilateral working groups (acid rain)
157

Saskatchewan. *See* federal-provincial
relations; Rafferty–Alameda dams
Sauvé, Jeanne 41, 42
Science Council of Canada 102, 142,
152
Seaborn, Blair 34, 41–2, 44–5, 46, 64–5,
111, 196, 197–8, 199, 203
Sewage Containment and Treatment
Program 216
Shamrock Summit 159, 162, 163
Shaw, Robert 40, 44, 52, 196
Sifton, Clifford 102
Silent Spring (Carson) 29, 103
Skagit Valley project (BC) 106, 129
Slater, Bob 73, 78, 153–3
SO_2 110, 114, 126, 138, 158–9, 166;
international agreements on 137
SO_2 30 Per Cent Club 45, 125, 132,
135–6, 132, 134, 135–7, 157–8
South Moresby Park 7, 45, 47, 64, 85,
111; budget, funding of 181, 185,
186, 187; description of 171; DOE

and decision on 175; and DOE internal differences 183–4; as element of DOE mandate 168–9; and ENGOs 174, 181, 182–3, 185; establishment of 175–80; federal-provincial negotiations over 173–5, 178; final stage of decision on 180–6; and Haida Nation 175, 179, 182–3 (logging blockade); key events in decision on 176–8; and land claims 179; local opposition to 175, 178–9; and logging 175, 178, 179, 181–6, 189; and media process 231–2; signing of 186

Soviet Eastern Bloc countries 136

St Clair River 221

St Basile-le-Grand PCB fire 14

Standing Committee on Fisheries and Forestry 165–6

State of Environment reports (DOE) 113

Ste Marie, Genevieve 27, 48–9, 112

Still Waters 166

Stockholm U.N. Conference on Human Settlements (1972) 40–1, 53, 54, 58, 64, 124, 141, 147, 232, 243

Strong, Maurice 40, 124, 141

sustainable-development concept 5, 14–15, 114, 115–16, 118–19, 125, 143, 235–7

Suzuki, David 175, 176, 180

Sweden 152; and 30 Per Cent Club 136

Switzerland 137–8

Tanker Safety and Pollution Prevention Convention 133

telecommunications sector 115

Thatcher, Margaret 158

thermodynamics, defined 8–9

Third World countries 139–40

30 Per Cent Club. *See* SO$_2$ 30 Per Cent Club

Thompson, Andy 19

Three-Mile Island 36

Time Lost: A Demand for Action on Acid Rain 166

Tinney, Roy 19

Tourism Canada 172

Toxic Chemicals Management Program 220

toxic substances: issues 55; as DOE priority 220; priority substances list 227; legislation on 216–18; regulation of 14, 32, 109, 118; working groups on 113–14, 116

Transport Canada 22

Treasury Board Secretariat: and DOE budget, expenditures 27, 47, 48, 81; and DOE creation 17, 18

Trudeau, Pierre Elliott 40, 42, 43, 44, 45, 46, 60, 62, 63, 201–2, 234; establishes DOE 3, 17, 19–20

Turner, John 46

U.S.–Canada relations. *See* Canada–U.S. relations

Union Carbide 221; *see also* Bhopal chemical disaster (India)

United Kingdom: and acid rain 135, 158; and Helsinki Protocol 136

United Nations: Educational, Scientific and Cultural Organization 132–3; Environmental Program 132; *see also* Economic Commission for Europe

uranium mining 97

Vander Zalm, Bill 177, 178, 185

VOCs (volatile organic compounds).
  See NOx and VOCs

Water Management Service 16
West Virginia 155
Western Diversification program 186
Wetherup, Danielle 135, 136
Whelan, Eugene 45
White Paper, environmental (draft)
  54

Wildlife Service 123; integration of
  with Lands Directorate 32–3
World Commission on Environment
  and Economy. *See* Brundtland
  Commission
World Conservation Strategy confer-
  ence (1980) 124, 141–2, 147, 232
World Meteorological Organization
  133
Wosnow, Ron 111